Spatial Big Data, BIM and advanced GIS for Smart Transformation

Spatial Big Data, BIM and advanced GIS for Smart Transformation: City, Infrastructure and Construction

Special Issue Editors

Sara Shirowzhan
Willie Tan
Samad M. E. Sepasgozar

MDPI • Basel • Beijing • Wuhan • Barcelona • Belgrade

Special Issue Editors

Sara Shirowzhan
The University of New
South Wales
Australia

Willie Tan
National University of Singapore
Singapore

Samad M. E. Sepasgozar
The University of New
South Wales
Australia

Editorial Office
MDPI
St. Alban-Anlage 66
4052 Basel, Switzerland

This is a reprint of articles from the Special Issue published online in the open access journal *ISPRS International Journal of Geo-Information* (ISSN 2220-9964) from 2019 to 2020 (available at: https: //www.mdpi.com/journal/ijgi/special_issues/BIM_GIS).

For citation purposes, cite each article independently as indicated on the article page online and as indicated below:

LastName, A.A.; LastName, B.B.; LastName, C.C. Article Title. *Journal Name* **Year**, *Article Number*, Page Range.

ISBN 978-3-03936-030-7 (Pbk)
ISBN 978-3-03936-031-4 (PDF)

Cover image courtesy of Sara Shirowzhan.

Contents

About the Special Issue Editors

Sara Shirowzhan is a lecturer of City Analytics at the Faculty of Built Environment. She completed her Ph.D. in Geomatics Engineering with the School of Civil and Environmental Engineering, University of New South Wales, Australia. Sara started teaching in built environment disciplines from 2007 and currently teaches GIS, BIM, and major project courses at undergrad and postgrad levels. She has supervised 22 M.S. students so far and has been involved in university and government funded research projects. She has developed her skills in advanced technologies for a sustainable and smart built environment such as laser scanning, nD BIM, nD GIS, GIS-based app and dashboard creation, digital twins, big data analysis, VR. and AR.

Willie Tan is a tenured Professor at the School of Design and Environment (SDE), National University of Singapore (NUS) and a Visiting Professor at the Southeast University. He was formerly Head of Department of Building and Vice Dean (Academic) of SDE. He is currently the Program Director of the M.S. (Project Management) program. He has chaired accreditation visit teams for the Project Management Institute in Asia. Dr. Tan is Editor of the World Scientific Series on the Built Environment, an editor of *Journal of Spatial Science*, and an Editorial Board member of many journals. He has published widely in geomatics and construction and authored several books on project finance, housing, and research methods. He has served as a consultant for organizations in many countries. He has also served as an expert witness and as Distinguished Chief Judge of the Built Environment Industry Asia Awards. Dr. Tan has won many teaching awards including the NUS Outstanding Educator Award (2013) and Honour Roll. He is a former Fellow of the NUS Teaching Academy.

Samad M. E. Sepasgozar is a senior lecture in construction and property management, and convener of the Smart City and Infrastructure Cluster UNSW, which provides leadership and collaborative opportunities in the smart city space. He is graduated from the School of Civil Engineering, the University of New South Wales, Australia, and published over 100 scholarly papers concerning on digital technology applications such as automation, geo spatial solutions, digital twin, deep learning, and big data analysis. Since his Ph.D., he has received a global and national award on his research output yearly, and received awards as the top reviewer and top five inspiring university teacher. He supervised over 50 masters' research project, and is currently supervising five Ph.D. students. He also serves as Editorial Board member of *Construction Innovation* journal, ranked Q1, and a Guest Editor of over 10 highly ranked Scopus journals.

Preface to "Spatial Big Data, BIM and advanced GIS for Smart Transformation: City, Infrastructure and Construction"

Shirowzhan et al. [1] discuss the value of digital twin and cybergis in improving connectivity and measuring the impact of infrastructure construction planning in smart cities. This chapter discusses selective technologies that can potentially contribute to developing an intelligent environment and smarter cities. Although the connectivity and efficiency of smart cities is important, the analysis of the impact of construction development and large projects in the city is crucial to decision and policy makers before the project is approved. This chapter refers to the need for advanced tools such as mobile scanners, geospatial artificial intelligence, unmanned aerial vehicles, geospatial augmented reality apps, light detection, and ranging in smart cities. In line with smart city technology development, this book includes 10 chapters covering trending topics, which are briefly mentioned in this preface.

Mendoza-Silva et al. [2] present a simulator for improving smart parking practices by modelling drivers with activity plans. This experimental study offers a parking occupancy simulator to support a smart system for managing parking. This paper is critical for use in extending smart city practices, as it shows how the process of developing a simulator assists in smart parking development from design to implementation.

Gu et al. [3] propose a bike optimization algorithm to increase the efficiency of bike stations and the sharing system in Shenzhen, China. Station-free bike sharing systems were recently introduced in China in line with smart city practices. They propose an optimization algorithm to match bike offers and rides.

Wu et al. [4] used an agent-based model simulation of human mobility with the use of mobile phone datasets and spatial big data analysis. They identified individual travel in urban areas and simulated commuting behaviors of residents using an agent-based model.

Rupi et al. [5] describe the use of numerical methods to match the network demand and supply of bicycles. This is a useful study for the improvement of city infrastructure using spatial data sets.

Li et al. [6] investigated the distribution of railways in China using indicators such as network density, proximity, travel time, train frequency, population, and gross domestic product (GDP). They then evaluated China's railway network distribution using geographic information system (GIS).

Dong et al. [7] present a novel algorithm of direction-aware continuous moving K-nearest neighbor queries in road networks. They show how object azimuth information can be used to determine the moving direction toward the query object.

Wang et al. [8] propose a hybrid framework for the high-performance modelling of 3D pipe networks. Three-dimensional modelling is a trending topic in smart city literature [9]. They explain how instantiation technology significantly improves the rendering performance of the 3D pipe networks.

Han et al. [9] present an efficient staged evacuation planning algorithm for multi-exit buildings. This algorithm can be tested using advanced big data simulations and virtual reality technologies.

Bibliography

[1] Shirowzhan, S.; Tan, W.; Sepasgozar, S.M.E. Digital Twin and CyberGIS for Improving Connectivity and Measuring the Impact of Infrastructure Construction Planning in Smart Cities. *ISPRS Int. J. Geo-Inf.* **2020**, *9*, 240.

[2] Silva, G.M.M.; Gould, M.; Montoliu, R.; Torres-Sospedra, J.; Huerta, J. An Occupancy Simulator for a Smart Parking System: Developmental Design and Experimental Considerations. *ISPRS Int. J. Geo Inf.* **2019**, 8, 212.

[3] Gu, Z.; Zhu, Y.; Zhang, Y.; Zhou, W.; Chen, Y. Heuristic Bike Optimization Algorithm to Improve Usage Efficiency of the Station-Free Bike Sharing System in Shenzhen, China. *ISPRS Int. J. Geo Inf.* **2019**, *8*, 239.

[4] Wu, H.; Liu, L.; Yu, Y.; Peng, Z.; Jiao, H.; Niu, Q. An Agent-based Model Simulation of Human Mobility Based on Mobile Phone Data: How Commuting Relates to Congestion. *ISPRS Int. J. Geo Inf.* **2019**, *8*, 313.

[5] Rupi, F.; Poliziani, C.; Schweizer, J. Data-driven Bicycle Network Analysis Based on Traditional Counting Methods and GPS Traces from Smartphone. *ISPRS Int. J. Geo Inf.* **2019**, *8*, 322.

[6] Li, M.; Guo, R.; Li, Y.; He, B.; Fan, Y. The Distribution Pattern of the Railway Network in China at the County Level. *ISPRS Int. J. Geo Inf.* **2019**, *8*, 336.

[7] Dong; Yuan; Shang, Y.; Ye; Zhang, L.; Tianyang, D.; Yuan, L.; Ye, Y. Direction-Aware Continuous Moving K-Nearest-Neighbor Query in Road Networks. *ISPRS Int. J. Geo Inf.* **2019**, *8*, 379.

[8] Wang, S.; Sun, Y.; Sun, Y.; Guan, Y.; Feng, Z.; Lu, H.; Cai, W.; Long, L. A Hybrid Framework for High-Performance Modeling of Three-Dimensional Pipe Networks. *ISPRS Int. J. Geo Inf.* **2019**, *8*, 441.

[9] Han, L.; Guo, H.; Zhang, H.; Kong, Q.; Zhang, A.; Gong, C. An Efficient Staged Evacuation Planning Algorithm Applied to Multi-Exit Buildings. *ISPRS Int. J. Geo-Inf.* **2020**, *9*, 46.

Sara Shirowzhan, Willie Tan, Samad M. E. Sepasgozar
Special Issue Editors

International Journal of
Geo-Information

Editorial

Digital Twin and CyberGIS for Improving Connectivity and Measuring the Impact of Infrastructure Construction Planning in Smart Cities

Sara Shirowzhan [1], Willie Tan [2] and Samad M. E. Sepasgozar [1,*]

[1] Faculty of Built Environment, University of New South Wales, Sydney, NSW 2052, Australia;
    s.shirowzhan@unsw.edu.au
[2] Department of Building, National University of Singapore, Lower Kent Ridge Road,
    Singapore 117566, Singapore; willietan@nus.edu.sg
* Correspondence: sepas@unsw.edu.au; Tel.: +61-469-628-400

Received: 7 April 2020; Accepted: 8 April 2020; Published: 12 April 2020

Abstract: Smart technologies are advancing, and smart cities can be made smarter by increasing the connectivity and interactions of humans, the environment, and smart devices. This paper discusses selective technologies that can potentially contribute to developing an intelligent environment and smarter cities. While the connectivity and efficiency of smart cities is important, the analysis of the impact of construction development and large projects in the city is crucial to decision and policy makers, before the project is approved. This raises the question of assessing the impact of a new infrastructure project on the community prior to its commencement—what type of technologies can potentially be used for creating a virtual representation of the city? How can a smart city be improved by utilizing these technologies? There are a wide range of technologies and applications available but understanding their function, interoperability, and compatibility with the community requires more discussion around system designs and architecture. These questions can be the basis of developing an agenda for further investigations. In particular, the need for advanced tools such as mobile scanners, Geospatial Artificial Intelligence, Unmanned Aerial Vehicles, Geospatial Augmented Reality apps, Light Detection, and Ranging in smart cities is discussed. In line with smart city technology development, this Special Issue includes eight accepted articles covering trending topics, which are briefly reviewed.

Keywords: digital twin; smart city; smart parking; GIS; lidar; point cloud; machine learning; point-based algorithms; mobile laser scanner; infrastructure construction; urban computing; CyberGIS; big data; artificial intelligence

1. Introduction

From the practical perspective, a smart city has the capability to capture real-time data that are communicated among stakeholders for optimizing decision-making by deploying artificial intelligence. This is achievable by making activities, services, and businesses smart, e.g., smart real estate, smart transportation, smart construction, smart healthcare system, smart building, smart home, smart transportation, and smart parking. For example, Virtual Singapore [1] is a dynamic 3D city model with a collaborative platform and data sharing system. This virtual city was initiated and funded by the National Research Foundation (NRF) with a $73 million investment.

Over the past year, the number of mobile users has increased by over two percent, up to 5.11 billion globally [2]. The number of internet users is also increasing. Due to the current Covid-19 outbreak, many people are working from home (WfH) and shop using online platforms. Geographic Information Science and System (GIS—Geographic Information Systems) technologies and IT services are used to analyse spatial and temporal data collected from various organisations to get better insights regarding current trends and model future trends and their impacts on communities and well-being. This big shift from traditional workplaces and shopping towards WfH and online shopping underscores the importance of further developing smart city infrastructures and deploying geospatial technologies to address future needs. This paper identifies selected trends in geospatial science, particularly the applications of GIS. In addition, this paper observes newly developed online apps such as ArcGIS Urban, used for predicting future impacts of developing urban areas in three dimensions. These technologies provide useful tools for smart city stakeholders and users to predict future implications of the proposed plans and collaborate with the organisations to achieve more appropriate outcomes, considering various criteria including sustainable development goals (SDGs) at various scales.

Digital twins of cities have recently attracted much attention as a useful virtual platform that captures changes to the physical environment in the city and all associated activities and movements [3]. Figure 1a–d schematically illustrate a digital–physical twin of a smart city, including the data management process and dashboard development. Figure 1e,f shows some examples developed by the first author. Using sensors, Unmanned Aerial Vehicles (UAVs), satellites, and different technologies, the physical entities, activities, behaviours, and interactions are required to be connected to a digital model [3] for a more realistic data platform. Integration of the digital twin as a 3D representation of the city and associated information can be used for the assessment of the performance of the city and selected construction projects using a data management system. Apps such as ArcGIS urban can also help us to evaluate the impact of a new project before it is implemented. Such digital twins, in conjunction with sensors and other advanced data collection technologies, can help in better modelling the strategic behaviours of agents [4].

This editorial is divided into two sections: (i) the development of advanced tools such as miniaturization of sensors and mobile scanners, geospatial AI, Unmanned Aerial Vehicles (UAVs), geospatial AR apps, and Light Detection and Ranging (Lidar); as well as (ii) applications of the tools in cities and products such as Self-Driving Vehicles and Smart Cities. Finally, the papers included in this Special Issue are reviewed.

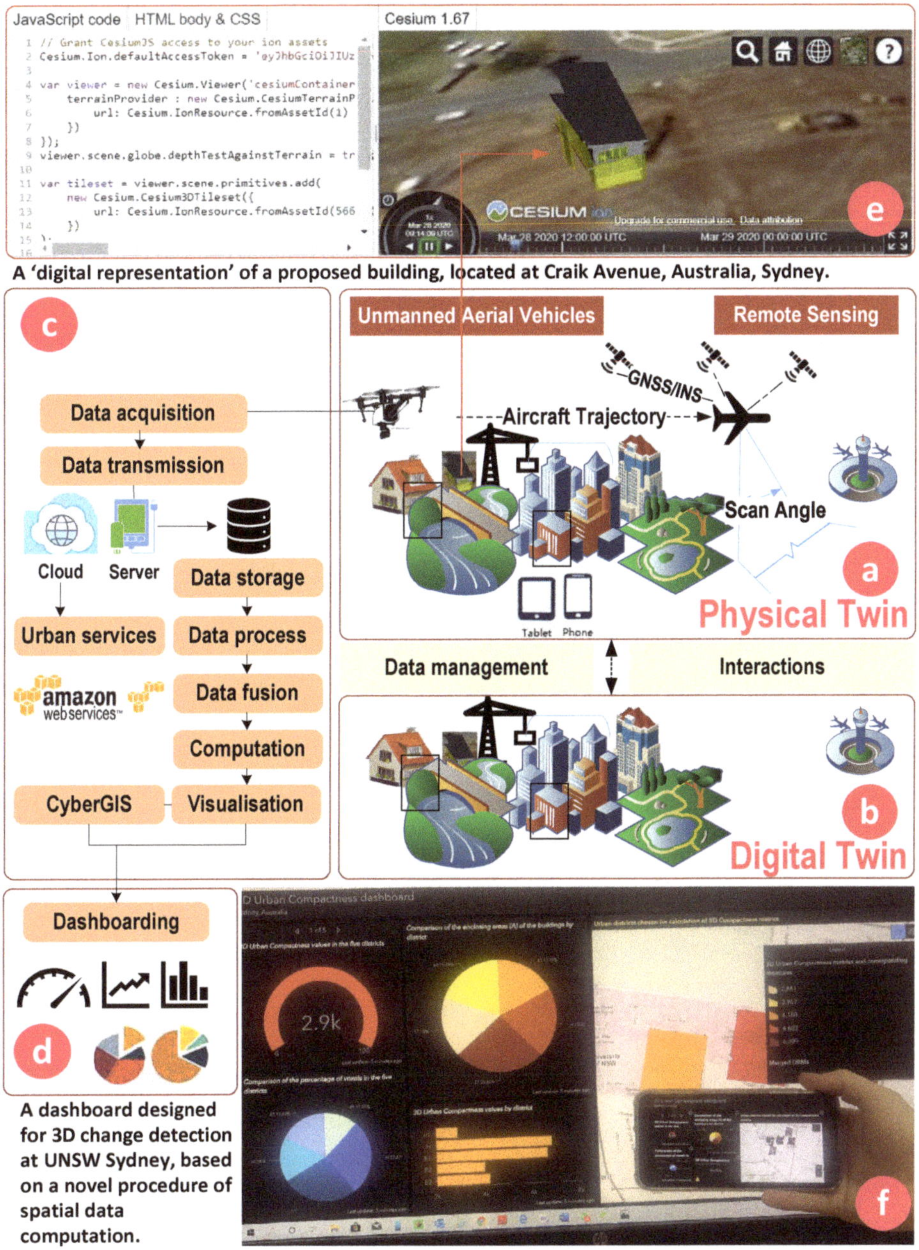

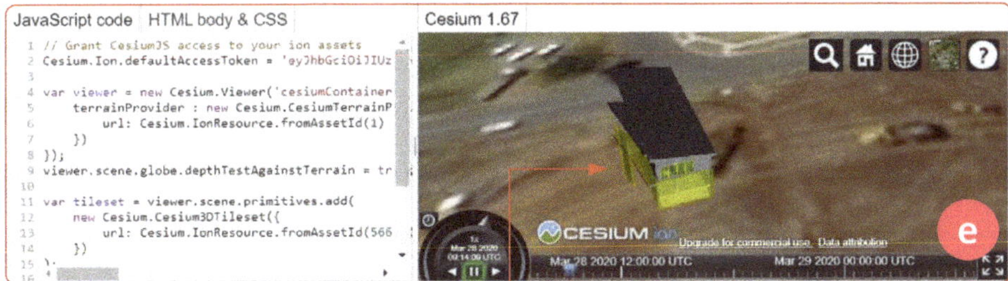

Figure 1. Demonstrating digital–physical twin at the city scale, (**a**) city physical twin; (**b**) city digital twin; (**c**) data management and interactions; (**d**) dashboard development based on computation; (**e**) integration of Building Information Modelling (BIM) and Geographic Information Systems (GIS) for a digital model of a proposed building; (**f**) dashboard developed by the first author used for making better insights from data.

2. Advanced Technologies

This section introduces the different tools and technologies that are critical to create a digital twin for a smart city. These critical technologies include network technology, sensors, artificial intelligence, big data, and Lidar technologies [5]. The integration of sensors with GIS in city analytics is a new geospatial trend that can significantly improve smart city technology. While the cost of utilizing such sensors is high, another emerging trend is to miniaturize sensors and data acquisition tools such as innovative bee-sized drones and mini satellites to generate useful data in an efficient and cost-effective manner. Selected technologies useful to improve the connectivity in smart cities which can be used for digital-physical twin are discussed as follows but can be further investigated in the future (See Figure 1).

2.1. CyberGIS

The combination of cyber infrastructure and GIS offers new capacity for online spatial analysis as a new generation of GIS. In practice, CyberGIS refers to the deployment of GIS on a web platform instead of running from a single desktop. CyberGIS Gateway, Toolkit, and Middleware can enable CyberGIS to offer open source, open access, and the possibility of integration [6–8]. CyberGIS analytics deal with interoperability of geospatial big data [9] and processing software programs [7]. The implementation of CyberGIS involves in many issues such as cloud base working, hardware/software challenges, internet speed, and the availability of skilled people to prepare, process, and use it. Advances in cloud computing, web mapping, and new algorithms for spatial data analysis in a quicker manner and including more dimensions of data (such as voxels instead of pixels) are required. CyberGIS also offers real-time accessibility to the outcome of computations, and dashboards to monitor changes and trends as well as gain insights from the available data connected to the dashboards. High-performance computing and networking (HPCN) (including CyberGIS) enables parallel processing used for running big GIS data and CyberGIS applications quickly and efficiently. Since big data is generated in different businesses, applying HPCN in GIS can be further investigated in different contexts.

2.2. Integration of GIS and BIM

While GIS is becoming more intelligent using cyberinfrastructure, its indoor applications also will be further developed. In this situation, Building Information Modelling (BIM), as a rich dataset for buildings including detailed information of the indoor built environment, will be helpful if fully available [10]. In this case, the integration of GIS and BIM will bring more advantages to the data sets including the combination of large- and small-scale built environments, looking at BIM models in a broader context of 3D geographic location, producing a more realistic model of built environments including buildings, vegetation, terrain, road network, and agents. For such integration, there are interoperability issues and factors, as Shirowzhan, Sepasgozar [11] discussed recently.

For real-time visualization and near real-time analysis of streaming data within GIS environments, technologies such as the Internet of Things are required to be interoperable with CyberGIS. The real-time output also depends on reliable "geospatial big data" computation algorithms to deal with volume, variety, and velocity of data [12–15]. There remains geospatial big data challenges that should be investigated in terms of availability of data on multi-cloud models, data integrity, data standards, and heterogeneity.

2.3. Laser Scanning Technologies

In recent years, laser scanning technology has been increasingly used for different purposes, including construction projects [16,17], for the provision of high-resolution point clouds of complex objects [18,19]. Another important application of point clouds is to track the physical progress of buildings and urban developments [20]. Progress tracking is the process of identifying differences and/or geometrical changes in an object over a specified period [21,22]. Monitoring physical progress in construction projects is crucial for measuring efficiency and productivity, but the current practice is cumbersome and complicated. In addition to these applications, laser scanners can be used to collect high point density data with a high level of accuracy, fine spatial resolution and in a shorter time for deformation detection [23,24]. For these applications, a wide range of algorithms and methods are utilized to optimize point cloud processing including filtering, simplification, feature recognition, segmentation, and registration [25]. Despite all these applications of point clouds, the use of laser scanning for deformation monitoring, 3D change detection, and volume estimation using bi-temporal datasets is still in infancy [20,26,27]. One of the challenges of deformation or change detection over time, on a regular basis, is related to handling the extremely large volume of point cloud data. Figure 2 illustrates a field point cloud data set for a light rail infrastructure. One can envision how big the data set will be if a practitioner continues the data collection task every day for a year or two. A handheld mobile scanner, Zeb Revo, was used to collect point cloud data on different occasions. As Sepasgozar, Forsythe [22] mentioned, the advantage of this type of mobile scanners is to collect data from surface objects easily and quickly.

Figure 2. Infrastructure construction survey using a Geo SLAM (Simultaneous Localisation And Mapping) hand-held scanner: (**a**) data acquisition from a light rail project in Randwick, Sydney; (**b**) illustration of tracing route in red.

2.4. Machine/Deep Learning Algorithms

Machine/deep learning algorithms are in use for processing various types of data, including images and point clouds. These algorithms are used for the detection of objects and changes. These algorithms have also been used for Artificial Intelligence and the detection of moving objects such as the identification of hats on/off workers in a construction site for monitoring safety in a workplace.

There are several algorithms including machine/deep learning methods available for processing Lidar point clouds and spatial data sets to achieve a reliable result for change detection of moving objects [25,28,29]. For example, Shirowzhan, Sepasgozar [25] applied machine learning and point-based algorithms to show 3D changes of urban environment over time using bi-temporal point clouds. Many algorithms have been proposed, such as the Iterative Closest Point (ICP) [30], Cloud to Cloud (C2C) method [31], and Multiscale Model to Model Cloud Comparison (M3C2) Lagüela, Díaz-Vilariño [32]. In M3C2, specifications of the point cloud, such as density, affect the choice of the main normal direction. If the changes are mainly in vertical dimension, such as in airborne point clouds, then the vertical normal should be chosen for the normal direction. However, the current algorithms need to be further extended and modified to be able to deal with spatiotemporal big data enabling three-dimensional modelling and visualization.

2.5. High Performance Computing

Recent publications tend to use spatiotemporal computing or different approaches of high-performance computing on modelling vector-borne disease transmission [33] and crime analysis [34]. Other studies also used digital twin applications to improve disaster management. Fan, Zhang [4] utilized a digital twin for use in disaster management using crisis informatics and information in a cyber infrastructure. They suggested a dynamic network analysis and a gaming approach for modelling behaviours of multi-actors and evaluating the performance of resilience and relief efforts.

3. Smart Cities Including Smart Elements

The technological part of a smart city refers to the use of information systems for planning, controlling, and managing critical infrastructure [35,36]. Recently, a wide range of tools and applications have been introduced to facilitate implementing smart cities and capturing user behaviour from different social and managerial perspectives [37]. A wider range of geo-spatial technologies are required for smart cities in order to generate data and monitor changes over time in the city. For example, hand-held mobile scanners can collect data on a daily basis and help practitioners to process changes over time (see Figure 2). The point clouds generated on a daily task can be communicated with other stakeholders and experts using cloud base platforms. However, challenges of big data analysis, interoperability, and implementations need to be investigated and resolved first. For digital twin applications in smart cities, agent's behaviour also need to be reflected and modelled within the geospatial platforms. There are still gaps in this domain, including modelling the dynamic behaviour of interacting subsystems suggested by Lom and Pribyl [38]. They also recommend investigating the behaviour of smart city agents and the identification of the details of subsystem behaviour as well as efficient ways of information exchange. In addition, for such applications, the interoperability of systems such as GIS with BIM and virtual/augmented reality applications is still challenging [11].

One challenge on the path towards smarter cities is to develop multi-criteria decision-making models and simulation scenarios. One of the recently developed applications by ESRI, i.e., ArcGIS Urban, is a type of appropriate application suitable for multi-criteria decision making in 3D urban environments. Nowadays, dashboards are being increasingly used for the provision of better insights for informed decision making in smart cities. These dashboards are connected to the sensors that provide data and using these dashboards experts can monitor current situations, changes and make smarter data driven decisions. Of course, these sensors produce big data sets that need to be stored and analyzed and cloud-based GIS platforms (refer to CyberGIS here) are most appropriate for dealing with such huge data sets.

Relevant and insightful case studies on CyberGIS implementation requires the identification of the challenges in a cloud-based working environment, hardware/software interoperability, speed of internet, ease-of-use application developments, and accessibility of the application to users.

There is also an urgent need to connect the online GIS tools to the supercomputing environment gateways for the projects with very big data sets. Advances in cloud computing, web mapping, and new algorithms for spatial 3D data processing and analysis using voxels are also required for acceleration of smart city developments considering more dimensions of urban developments.

An agenda for further investigation for smart cities and CyberGIS is to identify factors influencing the technology adoption process. There are several concepts which should be examined separately such as business readiness in utilizing new technologies, technology adoption [39], diffusion [40], technology dissemination [41], and implementation process [42,43]. This requires the development a taxonomy of technologies for the better understanding of different available technologies in smart cities. This practice has started in the construction field, where Sepasgozar and Davis [44] categorized technologies based on their adoption issues. Adopting this categorization, the useful technologies for smart cities are (i) network and office work technologies as highly penetrated information technologies (IT) and information and communication technologies (ICT) influencing human productivity, e-commerce, urban services [37], e-government [45], agriculture, and e-banking in smart cities. Recently, many communication technologies have arisen, such as social media [46], Skype, Teams, and Zooms which are known to be useful for working from home or working remotely which are critical for smarter cities; (ii) design technologies such as Building Information Modelling (BIM) [11,47], Geographic Information Systems (GIS), and CyberGIS [18,48]; (iii) sensing technologies such as wearable sensors [49], RFID [50,51], IoT sensors, real time locating systems (RTLS) [52], laser scanners [22], GPS, Radar, cameras for smart transportation, smart parking, and smart construction (job-site management, tracking materials, site management, physical progress monitoring, and productivity, safety, emission [53,54], security, and remote controlling devices and diagnostic systems attached or imbedded in heavy equipment such as Grader or Crane); (iv) production technologies such as 3D Printing [55,56]; and (v) virtual technologies such as mixed reality and digital twin.

4. Topics Covered in This Issue

While selected topics of trending technologies and emerging issues were briefly discussed in the previous sections, this section reviews some of related topics addressed in the Special Issue.

Mendoza-Silva, Gould [57] offer a simulator for improving the smart parking practices by modelling drivers with activity plans. This experimental study offers a parking occupancy simulator to support a smart system for managing parking. This paper is critical for use in extending smart city practices, as it shows how the process of developing a simulator assists in smart parking development from design to implementation.

Gu, Zhu [58] propose a bike optimization algorithm to increase the efficiency of bike stations and the sharing system in the case of Shenzhen in China. Station-free bike sharing systems were recently introduced in China in line with smart city practices. They propose an optimization algorithm to match bike offers and rides.

Wu, Liu [59] use an agent-based model simulation of human mobility with the use of mobile phone datasets and spatial big data analysis. They identify individual travels in urban areas and simulate commuting behaviours of residents using an agent-based model.

Rupi, Poliziani [60] describe the use of numerical methods to match the network demand and supply of bicycles. This is a useful study in the improvement of the city infrastructure using spatial data sets.

Li, Guo [61] investigate the distribution of railways in China using indicators such as network density, proximity, travel time, train frequency, population, and Gross Domestic Product (GDP). They then evaluated China's railway network distribution using GIS.

Dong, Yuan [62] present a novel algorithm of direction-aware continuous moving K-nearest neighbor queries in road networks. They showed how object azimuth information can be used to determine the moving direction towards the query object.

Wang, Sun [63] propose a hybrid framework for the high-performance modelling of 3D pipe networks. Three-dimensional modeling is a trending topic in smart city literature [25,64]. They explain how instantiation technology significantly improves the rendering performance of the 3D pipe networks.

Han et al. [65] present an efficient staged evacuation planning algorithm for multi-exit buildings. This algorithm can be tested using advanced big data simulations and virtual reality technologies.

Author Contributions: All authors involved in writing-review and editing. All authors have read and agreed to the published version of the manuscript.

Funding: This research received no external funding.

Conflicts of Interest: The authors declare no conflict of interest.

References

1. National Research Foundation (NRF). Virtual Singapore. 2020 28/03/2020. Available online: https://www.nrf.gov.sg/programmes/virtual-singapore (accessed on 28 March 2020).
2. Kemp, S. Digital 2019: Global Internet Use Accelerates. 2020. Available online: https://wearesocial.com/blog/2019/01/digital-2019-global-internet-use-accelerates (accessed on 28 March 2020).
3. Qi, Q.; Tao, F.; Hu, T.; Anwer, N.; Liu, A.; Wei, Y.; Wang, L.; Nee, A. Enabling technologies and tools for digital twin. *J. Manuf. Syst.* **2019**. [CrossRef]
4. Fan, C.; Zhang, C.; Yahja, A.; Mostafavi, A. Disaster City Digital Twin: A vision for integrating artificial and human intelligence for disaster management. *Int. J. Inf. Manag.* **2019**, 102049. [CrossRef]
5. Shirowzhan, S.; Sepasgozar, S.; Li, H.; Trinder, J. Spatial compactness metrics and Constrained Voxel Automata development for analyzing 3D densification and applying to point clouds: A synthetic review. *Autom. Constr.* **2018**, *96*, 236–249. [CrossRef]
6. Wang, S.; Liu, Y.Y.; Padmanabhan, A. Open cyberGIS software for geospatial research and education in the big data era. *SoftwareX* **2016**, *5*, 1–5. [CrossRef]
7. Liu, Y.Y.; Padmanabhan, A.; Wang, S. CyberGIS Gateway for enabling data-rich geospatial research and education. *Concurr. Comput. Pract. Exp.* **2014**, *27*, 395–407. [CrossRef]
8. Wang, S. A CyberGIS Framework for the Synthesis of Cyberinfrastructure, GIS, and Spatial Analysis. *Ann. Assoc. Am. Geogr.* **2010**, *100*, 535–557. [CrossRef]
9. Shirowzhan, S.; Lim, S.; Trinder, J.; Li, H.; Sepasgozar, S. Data mining for recognition of spatial distribution patterns of building heights using airborne lidar data. *Adv. Eng. Inform.* **2020**, *43*, 101033. [CrossRef]
10. Vilutiene, T.; Hosseini, M.R.; Pellicer, E.; Zavadskas, E.K. Advanced BIM Applications in the Construction Industry. *Adv. Civ. Eng.* **2019**, *2019*, 6356107. [CrossRef]
11. Shirowzhan, S.; Sepasgozar, S.M.; Edwards, D.J.; Li, H.; Wang, C. BIM compatibility and its differentiation with interoperability challenges as an innovation factor. *Autom. Constr.* **2020**, *112*, 103086. [CrossRef]
12. Kumar, B.; Pandey, G.; Lohani, B.; Misra, S.C. A multi-faceted CNN architecture for automatic classification of mobile LiDAR data and an algorithm to reproduce point cloud samples for enhanced training. *ISPRS J. Photogramm. Remote Sens.* **2019**, *147*, 80–89. [CrossRef]
13. Deng, X.; Liu, P.; Liu, X.; Wang, R.; Zhang, Y.; He, J.; Yao, Y. Geospatial Big Data: New Paradigm of Remote Sensing Applications. *IEEE J. Sel. Top. Appl. Earth Obs. Remote Sens.* **2019**, *12*, 3841–3851. [CrossRef]
14. Deibe, D.; Amor, M.; Doallo, R. Big Data Geospatial Processing for Massive Aerial LiDAR Datasets. *Remote Sens.* **2020**, *12*, 719. [CrossRef]
15. Barik, R.K.; Misra, C.; Lenka, R.K.; Dubey, H.; Mankodiya, K. Hybrid mist-cloud systems for large scale geospatial big data analytics and processing: Opportunities and challenges. *Arab. J. Geosci.* **2019**, *12*, 32. [CrossRef]
16. Louis, J.; Dunston, P.S. Methodology for Real-Time Monitoring of Construction Operations Using Finite State Machines and Discrete-Event Operation Models. *J. Constr. Eng. Manag.* **2017**, *143*, 04016106. [CrossRef]
17. Sepasgozar, S.; Wang, C.; Shirowzhan, S. Challenges and Opportunities for Implementation of Laser Scanners in Building Construction. In Proceedings of the 33rd International Symposium on Automation and Robotics in Construction (ISARC 2016), Auburn, AL, USA, 18–21 July 2016.
18. Shirowzhan, S.; Sepasgozar, S. Spatial Analysis Using Temporal Point Clouds in Advanced GIS: Methods for Ground Elevation Extraction in Slant Areas and Building Classifications. *ISPRS Int. J. Geo Inf.* **2019**, *8*, 120. [CrossRef]
19. Shirowzhan, S.; Lim, S.; Trinder, J. Enhanced Autocorrelation-Based Algorithms for Filtering Airborne Lidar Data over Urban Areas. *J. Surv. Eng.* **2016**, *142*, 04015008. [CrossRef]
20. Shirowzhan, S.; Sepasgozar, S.; Liu, C. Monitoring physical progress of indoor buildings using mobile and terrestrial point clouds. In Proceedings of the Construction Research Congress 2018, New Orleans, LA, USA, 2–4 April 2018.
21. Singh, A. Review Article Digital change detection techniques using remotely-sensed data. *Int. J. Remote Sens.* **1989**, *10*, 989–1003. [CrossRef]
22. Sepasgozar, S.; Forsythe, P.; Shirowzhan, S. Evaluation of Terrestrial and Mobile Scanner Technologies for Part-Built Information Modeling. *J. Constr. Eng. Manag.* **2018**, *144*, 04018110. [CrossRef]

23. Sepasgozar, S.; Lim, S.; Shirowzhan, S.; Kim, P.; Nadoushani, Z.M. Utilisation of a New Terrestrial Scanner for Reconstruction of As-built Models: A Comparative Study. In Proceedings of the ISARC International Symposium on Automation and Robotics in Construction, Oulu, Finland, 15–18 June 2015. Vilnius Gediminas Technical University, Department of Construction Economics & Property.
24. Sepasgozar, S.M.E.; Forsythe, P.J.; Shirowzhan, S. Scanners And Photography: A Combined Framework. In Proceedings of the 40th Australasian Universities Building Education Association (AUBEA) 2016 Conference, Cairns, Australia, 6–8 July 2016; Central Queensland University: Cairns, Australia.
25. Shirowzhan, S.; Sepasgozar, S.; Li, H.; Trinder, J.; Tang, P. Comparative analysis of machine learning and point-based algorithms for detecting 3D changes in buildings over time using bi-temporal lidar data. *Autom. Constr.* **2019**, *105*, 102841. [CrossRef]
26. Walton, G.; Delaloye, D.; Diederichs, M.S. Development of an elliptical fitting algorithm to improve change detection capabilities with applications for deformation monitoring in circular tunnels and shafts. *Tunn. Undergr. Space Technol.* **2014**, *43*, 336–349. [CrossRef]
27. Sepasgozaar, S.M.; Shirowzhan, S.; Wang, C. A Scanner Technology Acceptance Model for Construction Projects. *Procedia Eng.* **2017**, *180*, 1237–1246. [CrossRef]
28. Yan, Z.; Duckett, T.; Bellotto, N. Online learning for 3D LiDAR-based human detection: Experimental analysis of point cloud clustering and classification methods. *Auton. Robot.* **2019**, *44*, 147–164. [CrossRef]
29. Zhou, J.; Fu, X.; Zhou, S.; Zhou, J.; Ye, H.; Nguyen, H.T. Automated segmentation of soybean plants from 3D point cloud using machine learning. *Comput. Electron. Agric.* **2019**, *162*, 143–153. [CrossRef]
30. Faugeras, O.D.; Hebert, M.; Pauchon, E.; Ponce, J. Object representation, identification and Positioning from range data. In *Robotics and Artificial Intelligence*; Springer: Berlin/Heidelberg, Germany, 1984; pp. 255–277.
31. Lague, D.; Brodu, N.; Leroux, J. Accurate 3D comparison of complex topography with terrestrial laser scanner: Application to the Rangitikei canyon (N-Z). *ISPRS J. Photogramm. Remote Sens.* **2013**, *82*, 10–26. [CrossRef]
32. Lagüela, S.; Vilariño, L.D.; Martínez-Sánchez, J.; Armesto, J. Automatic thermographic and RGB texture of as-built BIM for energy rehabilitation purposes. *Autom. Constr.* **2013**, *31*, 230–240. [CrossRef]
33. Kang, J.-Y.; Aldstadt, J.; Vandewalle, R.; Yin, D.; Wang, S. A CyberGIS Approach to Spatiotemporally Explicit Uncertainty and Global Sensitivity Analysis for Agent-Based Modeling of Vector-Borne Disease Transmission. *Ann. Am. Assoc. Geogr.* **2020**, 1–19. [CrossRef]
34. Ajayakumar, J.; Shook, E. Leveraging parallel spatio-temporal computing for crime analysis in large datasets: Analyzing trends in near-repeat phenomenon of crime in cities. *Int. J. Geogr. Inf. Sci.* **2020**, 1–25. [CrossRef]
35. Washburn, D.; Sindhu, U. *Helping CIOs Understand "Smart City" Initiatives*; Forrester Research, Inc.: Cambridge, MA, USA, 2010.
36. Albino, V.; Berardi, U.; Dangelico, R.M. Smart Cities: Definitions, Dimensions, Performance, and Initiatives. *J. Urban Technol.* **2015**, *22*, 3–21. [CrossRef]
37. Sepasgozar, S.; Hawken, S.; Sargolzaei, S.; Foroozanfa, M. Implementing citizen centric technology in developing smart cities: A model for predicting the acceptance of urban technologies. *Technol. Forecast. Soc. Chang.* **2019**, *142*, 105–116. [CrossRef]
38. Lom, M.; Pribyl, O. Smart city model based on systems theory. *Int. J. Inf. Manag.* **2020**, 102092. [CrossRef]
39. Sepasgozar, S.; Davis, S.; Loosemore, M.; Bernold, L. An investigation of modern building equipment technology adoption in the Australian construction industry. *Eng. Constr. Arch. Manag.* **2018**, *25*, 1075–1091. [CrossRef]
40. Rogers, E.M. *Diffusion of Innovations*; Simon and Schuster: New York, NY, USA, 2010.
41. Sepasgozar, S.; Davis, S.; Loosemore, M. Dissemination Practices of Construction Sites' Technology Vendors in Technology Exhibitions. *J. Manag. Eng.* **2018**, *34*, 04018038. [CrossRef]
42. Foroozanfar, M.; Sepasgozar, S. Modeling Green Digital Technology Implementation in Construction. In Proceedings of the Construction Research Congress 2018, New Orleans, LA, USA, 2–4 April 2018.
43. Sepasgozar, S.; Davis, S.; Li, H.; Luo, X. Modeling the Implementation Process for New Construction Technologies: Thematic Analysis Based on Australian and U.S. Practices. *J. Manag. Eng.* **2018**, *34*, 05018005. [CrossRef]

44. Sepasgozar, S.; Davis, S. Digital Construction Technology and Job-site Equipment Demonstration: Modelling Relationship Strategies for Technology Adoption. *Buildings* **2019**, *9*, 158. [CrossRef]

45. Anand, A.; Vaidya, S.D.; Sharahiley, S.M. Role of integration in scaling of an e-Government project. *Transform. Gov. People Process. Policy* **2020**, *14*, 65–80. [CrossRef]

46. Laurell, C.; Sandström, C.; Berthold, A.; Larsson, D. Exploring barriers to adoption of Virtual Reality through Social Media Analytics and Machine Learning—An assessment of technology, network, price and trialability. *J. Bus. Res.* **2019**, *100*, 469–474. [CrossRef]

47. Pishdad-Bozorgi, P. Future Smart Facilities: State-of-the-Art BIM-Enabled Facility Management. *J. Constr. Eng. Manag.* **2017**, *143*, 02517006. [CrossRef]

48. Shirowzhan, S.; Sepasgozar SM, E.; Zaini, I.; Wang, C. An integrated GIS and Wi-Fi based Locating system for improving construction labor communications. In Proceedings of the 34th International Symposium on Automation and Robotics in Construction, Taipei, Taiwan, 28 June–1 July 2017.

49. Awolusi, I.; Marks, E.D.; Hallowell, M. Wearable technology for personalized construction safety monitoring and trending: Review of applicable devices. *Autom. Constr.* **2018**, *85*, 96–106. [CrossRef]

50. Fu, H.-P.; Chang, T.-H.; Lin, A.; Du, Z.-J.; Hsu, K.-Y. Key factors for the adoption of RFID in the logistics industry in Taiwan. *Int. J. Logist. Manag.* **2015**, *26*, 61–81. [CrossRef]

51. Wang, H.; He, J.; Pan, Y. A Solution Framework in Traffic Congestion Management Using RFID Technology. In Proceedings of the 2015 International Industrial Informatics and Computer Engineering Conference, Xi'an, China, 10–11 January 2015; Atlantis Press: Paris, France, 2015.

52. Li, H.; Chan, G.; Wong, K.W.; Skitmore, M. Real-time locating systems applications in construction. *Autom. Constr.* **2016**, *63*, 37–47. [CrossRef]

53. Sepasgozar, S.; Li, H.; Shirowzhan, S.; Tam, V.W.Y. Methods for monitoring construction off-road vehicle emissions: A critical review for identifying deficiencies and directions. *Environ. Sci. Pollut. Res.* **2019**, *26*, 15779–15794. [CrossRef] [PubMed]

54. Sepasgozar, S.M.E.; Blair, J. Measuring non-road diesel emissions in the construction industry: A synopsis of the literature. *Int. J. Constr. Manag.* **2019**, 1–16. [CrossRef]

55. Tahmasebinia, F.; Niemelä, M.; Ebrahimzadeh, S.; Lai, T.; Su, W.; Reddy, K.R.; Shirowzhan, S.; Sepasgozar, S.; Marroquin, F.A. Three-Dimensional Printing Using Recycled High-Density Polyethylene: Technological Challenges and Future Directions for Construction. *Buildings* **2018**, *8*, 165. [CrossRef]

56. Tahmasebinia, F.; Sepasgozar, S.M.; Shirowzhan, S.; Niemela, M.; Tripp, A.; Nagabhyrava, S.; Ko, Z.M.K.; Alonso-Marroquin, F. Criteria development for sustainable construction manufacturing in Construction Industry 4.0. *Constr. Innov.* **2020**. [CrossRef]

57. Silva, G.M.M.; Gould, M.; Montoliu, R.; Torres-Sospedra, J.; Huerta, J. An Occupancy Simulator for a Smart Parking System: Developmental Design and Experimental Considerations. *ISPRS Int. J. Geo Inf.* **2019**, *8*, 212. [CrossRef]

58. Gu, Z.; Zhu, Y.; Zhang, Y.; Zhou, W.; Chen, Y. Heuristic Bike Optimization Algorithm to Improve Usage Efficiency of the Station-Free Bike Sharing System in Shenzhen, China. *ISPRS Int. J. Geo Inf.* **2019**, *8*, 239. [CrossRef]

59. Wu, H.; Liu, L.; Yu, Y.; Peng, Z.; Jiao, H.; Niu, Q. An Agent-based Model Simulation of Human Mobility Based on Mobile Phone Data: How Commuting Relates to Congestion. *ISPRS Int. J. Geo Inf.* **2019**, *8*, 313. [CrossRef]

60. Rupi, F.; Poliziani, C.; Schweizer, J. Data-driven Bicycle Network Analysis Based on Traditional Counting Methods and GPS Traces from Smartphone. *ISPRS Int. J. Geo Inf.* **2019**, *8*, 322. [CrossRef]

61. Li, M.; Guo, R.; Li, Y.; He, B.; Fan, Y. The Distribution Pattern of the Railway Network in China at the County Level. *ISPRS Int. J. Geo Inf.* **2019**, *8*, 336. [CrossRef]

62. Dong, T.; Yuan, L.; Shang, Y.; Ye, Y.; Zhang, L. Direction-Aware Continuous Moving K-Nearest-Neighbor Query in Road Networks. *ISPRS Int. J. Geo Inf.* **2019**, *8*, 379. [CrossRef]

63. Wang, S.; Sun, Y.; Sun, Y.; Guan, Y.; Feng, Z.; Lu, H.; Cai, W.; Long, L. A Hybrid Framework for High-Performance Modeling of Three-Dimensional Pipe Networks. *ISPRS Int. J. Geo Inf.* **2019**, *8*, 441. [CrossRef]

64. Shirowzhan, S.; Trinder, J.; Osmond, P. New Metrics for Spatial and Temporal 3D Urban Form Sustainability Assessment Using Time Series Lidar Point Clouds and Advanced GIS Techniques. In *Urban Design*; IntechOpen: London, UK, 2019.
65. Han, L.; Guo, H.; Zhang, H.; Kong, Q.; Zhang, A.; Gong, C. An Efficient Staged Evacuation Planning Algorithm Applied to Multi-Exit Buildings. *ISPRS Int. J. Geo-Inf.* **2020**, *9*, 46. [CrossRef]

 International Journal of
Geo-Information

Article

An Efficient Staged Evacuation Planning Algorithm Applied to Multi-Exit Buildings

Litao Han [1,2,*], Huan Guo [1], Haisi Zhang [1], Qiaoli Kong [1], Aiguo Zhang [3] and Cheng Gong [1]

1   College of Geomatics, Shandong University of Science and Technology, Qingdao 266590, China; 18765330191@163.com (H.G.); haiszu@163.com (H.Z.); qiaolikong@sdust.edu.cn (Q.K.); 18380590702@163.com (C.G.)
2   Key Laboratory of Geomatics and Digital Technology of Shandong Province, Shandong University of Science and Technology, Qingdao 266590, China
3   College of Computer and Information Engineering, Xiamen Institute of Technology, Xiamen 361024, China; zhangaiguo@xmut.edu.cn
*   Correspondence: hanlitao@sdust.edu.cn; Tel.: +86-15969852846

Received: 4 December 2019; Accepted: 13 January 2020; Published: 15 January 2020

Abstract: When the occupant density of buildings is large enough, evacuees are prone to congestion during emergency evacuation, which leads to the extension of the overall escape time. Especially for multi-exit buildings, it's a challenging problem to afford an effective evacuation plan. In this paper, a novel evacuation planning algorithm applied to multi-exit buildings is proposed, which is based on an indoor route network model. Firstly, evacuees are grouped by their location proximity, then all groups are approximately equally classified into several evacuation zones, each of which has only one safe exit. After that, all evacuation groups in the same zone are sorted by their shortest path length, then the time window of each evacuation group occupying the safe exit is calculated in turn. In the case of congestion at the safe exit, the departure time of each evacuation group is delayed in its arrival order. The objectives of the proposed algorithm include minimizing the total evacuation time of all evacuees, the travel time of each evacuee, avoiding traffic congestion, balancing traffic loads among different exits, and achieving high computational efficiency. Case studies are conducted to examine the performance of our algorithm. The influences of group number, group size, evacuation speed on the total evacuation time are discussed on a single-exit network, and that of partitioning methods and evacuation density on the performance and applicability in different congestion levels are also discussed on a multi-exit network. Results demonstrate that our algorithm has a higher efficiency and performs better for evacuations with a large occupant density.

Keywords: emergency evacuation; indoor route network; multi-exit buildings; staged evacuation; congestion

---

1. Introduction

With the rapid development of urban construction and building technology, more and more large buildings have been built in cities, and their internal structures are increasingly complex. When an emergency or disaster occurs inside the buildings, their complex internal structure makes it difficult for indoor occupants to evacuate as quickly as a disaster occurs outside, which leads to frequent tragedies. It is critical for emergency rescuers and evacuees to plan an effective emergency evacuation plan [1]. The reason for this is that the plan can not only provide a reasonable escape path for evacuees in the event of a disaster, but also provide a basis for rescuers to make a rescue plan. Additionally, it can also provide reasonable suggestions for the layout of fire control facilities and the design of escape routes inside the buildings [2,3].

Due to the rapid occurrence and spread of disasters, it is necessary to make the best emergency evacuation plan in the shortest time. Therefore, two key objectives of a practical evacuation plan are to ensure the shortest overall escape time and to design the plan as quickly as possible. So far evacuation plans can be roughly classified as optimization-oriented and simulation-oriented [2]. Our research belongs to the former category and aims at developing an optimal method to design evacuation plans. In this paper, we deeply analyze the staged-evacuation process in crowded indoor environments and present a simple and efficient algorithm for staged-evacuation path planning that is able to cope with multi-exit networks. Generally, indoor evacuation is a multi-exit evacuation problem. The algorithm first transforms the multi-exit evacuation problem into a single-exit problem by balancing the loads of evacuees at all emergency exits, then performs the single-exit evacuation. Our contribution includes: 1) for multi-exit indoor evacuation, a partitioned and staged evacuation planning approach is proposed, which effectively realizes the transform above and simplifies the planning of multi-exit evacuation; 2) for single-exit indoor evacuation, a new idea of determining the escape sequence of evacuees according to their shortest path length is proposed and verified, which improves the efficiency of developing evacuation plan. Furthermore, the efficient single-exit evacuation will effectively improve the efficiency of the multi-exit evacuation, because the multi-exit evacuation is composed of multiple single-exit evacuations in this paper.

The remainder of this paper is organized as follows. Section 2 reviews related work. Section 3 describes the problem. Section 4 gives related definitions and theorems and presents our method. Section 5 illustrates the results of the algorithm, evaluates its performance and effectiveness by a series of tests, and gives a testing simulation. Section 6 concludes the paper.

2. Related Work

From the perspective of implementation, Li et al. classify evacuation plans into two major types: spontaneous evacuation plans and organized evacuation plans [2]. The former is carried out by controlling the evacuation infrastructure (e.g., fire emergency lighting and dispersal indicator) while evacuees move spontaneously but are guided by the infrastructure. The latter is realized by controlling evacuees including their departure time, routes to safe exits, and so on. Each type of evacuation plans is applicable to a particular scenario.

Regarding spontaneous evacuation plans, many simulation models have been put forward to analyze the significant factors or parameters that influence the evacuation process or can be used in evaluating the evacuation performance under different scenarios and strategies. Existing typical models include network flow based models [4], cellular-automata (CA) models [5,6], agent-based models [7–9], social-force models [10,11], lattice gas (LG) models [12,13], and so on. These models have been successfully applied to study crowd evacuation under various situations because of their great ability in representing some key elements influencing human behaviors during evacuation process, such as the impact of the occupant density around exits [14,15] and spatial distance on human behaviors. Flow based models are easy to construct while they lack social interaction between evacuees, human behavior in emergency conditions and hazards representation [4]. CA models are very flexible and effective in simulating evacuation process under complex environment while in contrast with multi-agent system, they have more primitive agents that are arranged on a rigid grid and interacting with each other by very simple rules. LG models present a special case of cellular automata modes that utilize biased random rules to simulate counter flow in channels, or to evaluate the impact of building parameters to the evacuation efficiency. Agent-based or multi-agent-based models can represent various types of agents with different attributes and their interactions are more complex [8], and the disadvantages of them are generally more computationally expensive than cellular automata. Social force models are a kind of continuous model applying Newton's second law to simulate pedestrian evacuation and are good at modeling interactions among pedestrians, but have low computational efficiency in simulating evacuation in complex buildings [6].

This paper focuses on organized evacuation planning requiring a comprehensive and effective escape plan for particular evacuation objectives according to different escape environments. Generally, these objectives include reducing traffic conflicts and minimizing the whole clearance time of all evacuees or the evacuation time of each evacuee. Network flow models, such as maximum-flow models and minimum-cost flow models [3,16–18], are the most widely used in optimizing the flow of evacuees, but they target the whole network and attempt to organize the origin, destination, and routes of evacuation flow at a mesoscopic level. The integer programming or linear programming method [16,19], as an exact algorithm, is applicable to small-scale problems and usually requires additional parameters (e.g., lower or upper bounds, etc.) that are generally difficult to estimate in advance. For large-scale evacuations, heuristic methods and scheduling algorithms are often adopted. The former, such as evolutionary algorithms [20] and ant colony optimization [21–23], are limited in terms of the quality of solutions and computing time. The latter are generally exact methods and are used to integrate the objectives and the constraints into the design of algorithms [2,24–26].

In the process of evacuation, if there is congestion, it will inevitably lead to the decrease of escape efficiency and even trample accident. In order to avoid congestion, waiting is necessary [27]. There are two ways of waiting in the strategies of scheduling algorithms. One is waiting at the starting point and the other is waiting on the way. Li et al. proposed an innovative method to make a staged-evacuation plan for emergency situations, but it is only applied to the network with a single safe exit and it is assumed that the speed of evacuees is constant and equal [25]. Later, they extended the staged-evacuation plan method from two aspects. One is to make it apply to multi-exit evacuation based on the time-extended network model by balancing traffic loads to different exits [26]. The other is to make it fit multi-speed evacuations with a single exit [2]. Although these algorithms get excellent results, iterative computation of the time-extended network results in their low efficiency.

It should be noted that another research direction closely related to indoor emergency evacuation is to dynamically plan an indoor evacuation path based on the real-time perceived situation information about the spread of a disaster [28–31]. However, so far, the research results in this direction are more applicable to the situation without indoor congestion. Additionally, the acquisition technology of real-time disaster environmental information in the case of fire has made great progress, but remains a challenging work. Encouragingly, the arrival of smart city offers real-time access to indoor evacuation information such as the distribution of evacuees and the development of an indoor disaster, which provides a data base for the real-time design of an evacuation scheme. This is one of the reasons why we pay attention to the efficiency of the algorithm.

3. Problem Description

Once an emergency occurs in a building with multiple exits, evacuees would choose the nearest exit to escape if there are a few occupants in the building. However, if there are more occupants, due to the limitation of the capacity of the escape path, it is prone for evacuees to congest at the corners or intersections of the path or safety exits, which reduces their escape speed, prolongs the overall evacuation time, and increases the probability of risk for them [25]. Therefore, the problem of indoor emergency evacuation studied in this paper is how to let all evacuees escape from the dangerous buildings with multiple safe exits in the shortest time when the capacity of indoor route is limited and congestion may occur during evacuation. Figure 1 shows the abstract representation of the studied evacuation problem. There are three safety exits namely E1, E2 and E3 in the indoor route network whose edges contain both path cost and capacity. *Ai* represents the room node. In this paper, it is assumed that the capacities of all locations in the route network are equal.

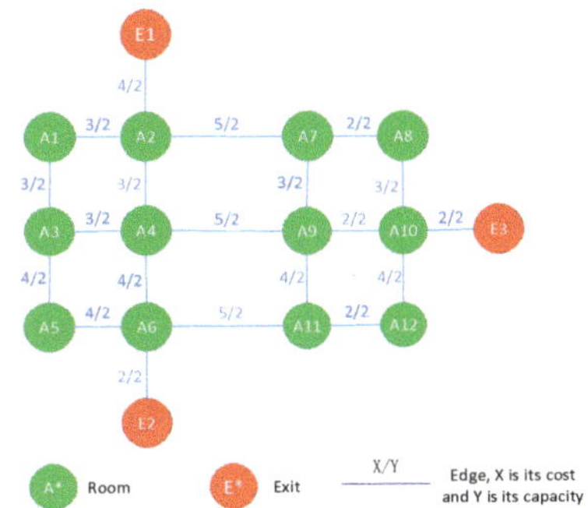

Figure 1. The indoor network with multiple exits.

4. Methodology

The purpose of emergency evacuation planning is to allocate a departure time and an escape path for each evacuee to ensure that all indoor occupants can safely and orderly escape to the safe area in the shortest time. When the number of evacuees is large, it is usually more effective to group them by the spatial proximity of their positions and then evacuate them in groups. A key issue during this procedure is congestion. To avoid the problem, two strategies are often adopted in case of emergency [25]. One such strategy is staged evacuation, and the other is simultaneous evacuation. However, in the second case, it is hard for subsequent groups to wait to escape until all prior groups have fully passed. Someone would likely abandon waiting at the congestion and escape blindly, which increases the degree of congestion and the total escape time. Therefore, we choose the former for emergency escape planning.

For an uncrowded multi-exit building, each safety exit has its corresponding service zone where evacuees can flee to the safety exit by their shortest paths [32]. When an emergency occurs, indoor occupants can escape from the exit of the evacuation zone where they are located. Therefore, the emergency evacuation planning for a multi-exit indoor emergency can be easily transformed into that for a single-exit indoor emergency according to the service zones of building exits (Figure 2). However, it must be noted that the goal of evacuation planning is to ensure the shortest overall evacuation time for all evacuees, rather than the shortest escape time for a single person. When the density of indoor occupants is very large or the distribution of them is non-uniform, the two factors should be taken into account when zoning. Only in this way can the number of evacuees in each evacuation zone be approximately equal, thereby making full use of all safety exits and getting the minimum of the total evacuation time.

Our proposed approach is mainly inspired by that in [25] which is only suitable for the single-exit problem. But our approach can solve the multi-exit problem well, especially with crowded evacuees. Its key issues include how to transform the multi-exit problem into the single-exit problem to make the total evacuation time minimum and how to improve the approach in [25] to get higher efficiency. Firstly, evacuees are grouped by their location proximity, then all groups are approximately equally classified into several evacuation zones by the improved Dijkstra algorithm according to the load of each exit. Secondly, all evacuation groups in the same zone are sorted by their shortest path length, then the time window of each evacuation group occupying the safe exit is calculated in turn. In the case of congestion at the safe exit, the departure time of each evacuation group is delayed in its arrival order.

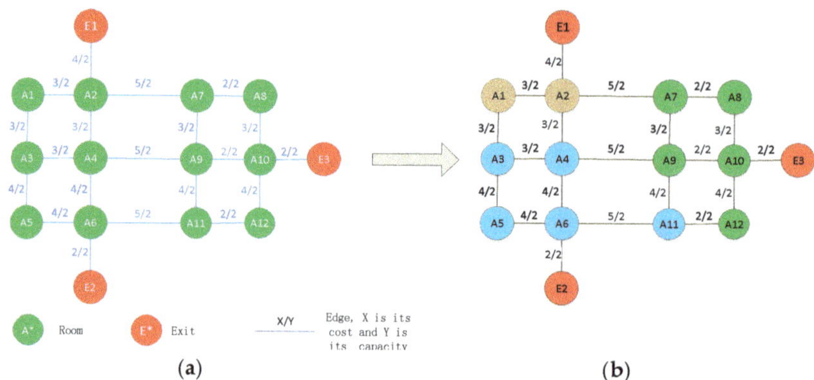

(a)   (b)

Figure 2. Transformation from the multi-exit route network to several single-exit route networks. (**a**) The multi-exit route network; (**b**) The partitioned multi-exit route network.

4.1. Staged Evacuation for Single-Exit Network

In the case of congestion, the evacuation process involves many factors, such as the weight and capacity of route network, the total number and total evacuation time of evacuees, the time when an evacuation order is issued, the waiting time, departure time and escape speed of each evacuee, etc. To describe and analyze the evacuation conveniently in theory, related variables are defined below.

n: the number of evacuation groups.

t_0: the earliest departure instant, that is, the time when an evacuation order is issued.

t_p^i: the time consumed by the escape group i from the origin to the safety exit E along the escape path.

t_e^i: the time consumed by the queue of group i to completely pass through a point such as E on the route network.

t_d^i: the delay of the departure time of the escape group i.

t_l^i: the time interval between the group i and its prior group along the route.

V: the escape speed of escape groups.

T: the total evacuation time of all escape groups that is the time from t_0 to the instant when the last evacuation group has passed through the emergency exit.

It is assumed that the escape speed V of each group is the same and the evacuation network has only one safety exit. At the same time, the staged evacuation process has four assumptions:

1. Each edge of the evacuation network has the same capacity, and when a node of the network is occupied by a group, other escape groups cannot pass through the node.
2. The delay time between two adjacent groups is to ensure that their time windows don't overlap or separate.
3. During the staged escape procedure, each group escapes along the shortest path to the exit.
4. The evacuation network is an undirected graph in which it will take the same cost to go or come along the same edge.

In the process of staged evacuation, each group arrives at the safety exit along their shortest path without any congestion. Then each group successively passes through the exit to complete the evacuation [25]. Accordingly, the total evacuation time of each group may be divided into three parts that include t_e^i, t_l^i and t_p^i. The total evacuation time T is equal to the time when the safety exit is occupied in the whole evacuation process, therefore T can be expressed as follows:

$$T = t_p^1 + \sum_{1}^{n} (t_l^i + t_e^i) \tag{1}$$

In the staged evacuation planning, the first key work is to determine the evacuation order of each evacuation group. We deeply analyzed the operation of the staged evacuation process and obtained Theorem 1. It is the basis of our proposed approach.

Theorem 1. *In order to obtain the shortest total evacuation time, all escape groups should escape in stages according to their distance from the emergency exit. The group near the exit has priority to depart and that far from the exit will be delayed if there are conflicts between their time windows.*

Proof of Theorem 1. Assuming that group Gi and group $Gi + 1$ need to be evacuated and only one of them is allowed to pass in the process of evacuation for the limits of path and node capacity. That is, when one group is passing through exit E, other groups can't pass through it. Then, several situations may occur at exit E (or path intersection) when the two groups are evacuated.

Situation 1. The two groups depart at the same time and arrive at the emergency exit successively without congestion.

As shown in Figure 3, assuming that Gi arrives at the emergency exit E before $Gi + 1$, the condition for no congestion at the exit is as follows:

$$t_p^i + t_e^i < t_p^{i+1} \tag{2}$$

At this time, $Gi + 1$ need not delay its departure time, $t_d^{i+1} = 0$. Their total evacuation time passing through the exit E successively can be expressed as follows:

$$T = t_p^i + t_e^i + t_l^{i+1} + t_e^{i+1} \tag{3}$$

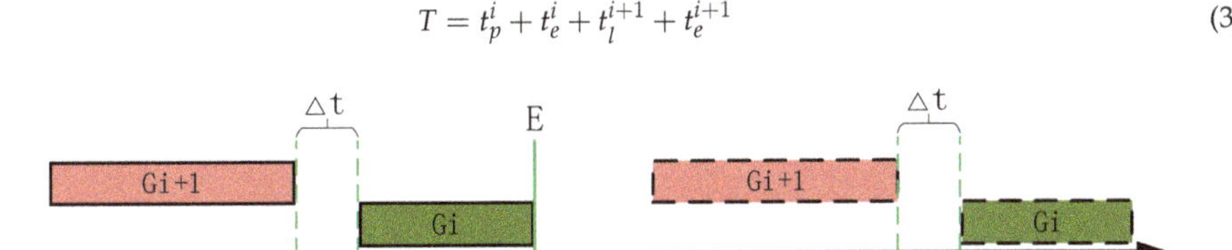

Figure 3. Two groups arrive at the exit successively without congestion.

Situation 2. Two groups depart at the same time and arrive at the exit at the same time, causing congestion

Group Gi and group $Gi + 1$ reach the emergency exit at the same time, namely $t_p^i = t_p^{i+1}$, as shown in Figure 4. In order to avoid the congestion of Gi and $Gi + 1$ at the emergency exit, one of them can start to escape immediately while the other must delay its departure time and wait at origin. To find the shortest total evacuation time, the delay time should ensure that the two groups pass through the emergency exit successively, and at the same time there is no time interval when they pass through the exit. There are two evacuation solutions to be discussed:

1. In case of emergency, Gi departs immediately and $Gi + 1$ delays its departure time. For $t_l^{i+1} = 0$, their total evacuation time is as follows

$$T = t_p^i + t_e^i + t_e^{i+1} \tag{4}$$

2. In case of emergency, $Gi + 1$ departs immediately and Gi delays its departure time. For $t_l^i = 0$, their total evacuation time is as follows:

$$T = t_p^{i+1} + t_e^{i+1} + t_e^i \tag{5}$$

because $t_p^i = t_p^{i+1}$, any of them can be delayed reasonably to avoid congestion when they arrive at the exit at the same time.

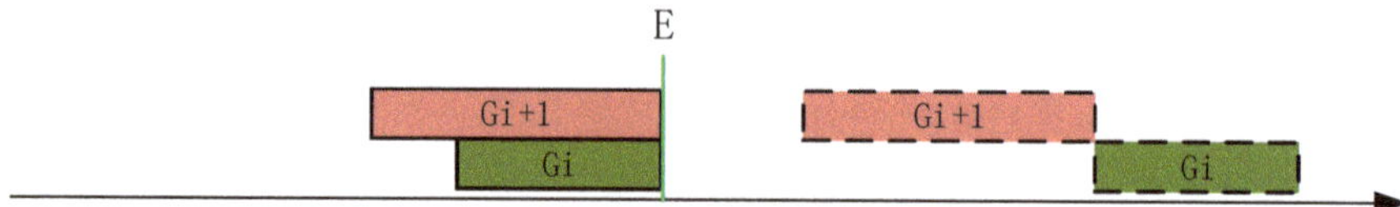

Figure 4. Two groups arrive at the emergency exit at the same time.

Situation 3. Two groups set out at the same time and arrive at the exit successively, causing congestion.

Assuming that Gi reaches E before $Gi + 1$, and $Gi + 1$ reaches E when Gi does not pass through the exit completely, as shown in Figure 5, the conditions for congestion of the two groups are $t_p^i < t_p^{i+1}$ and $t_p^i + t_e^i > t_p^{i+1}$. To find the shortest total evacuation time, the delay time should ensure that the two groups pass through the emergency exit successively, and at the same time there is no time interval when they pass through the exit. There are two evacuation solutions to be discussed:

1. In case of emergency, Gi departs immediately and $Gi + 1$ delays its departure time. For $t_l^{i+1} = 0$, their total evacuation time is as follows

$$Ta = t_p^i + t_e^i + t_e^{i+1} \tag{6}$$

2. In case of emergency, $Gi + 1$ departs immediately and Gi delays its departure time. For $t_l^{i+1} = 0$, their total evacuation time is as follows:

$$Tb = t_p^{i+1} + t_e^{i+1} + t_e^i \tag{7}$$

because $t_p^i < t_p^{i+1}$, $Ta < Tb$. Therefore, when two evacuation groups are congested, the group close to the emergency exit E should first start to escape.

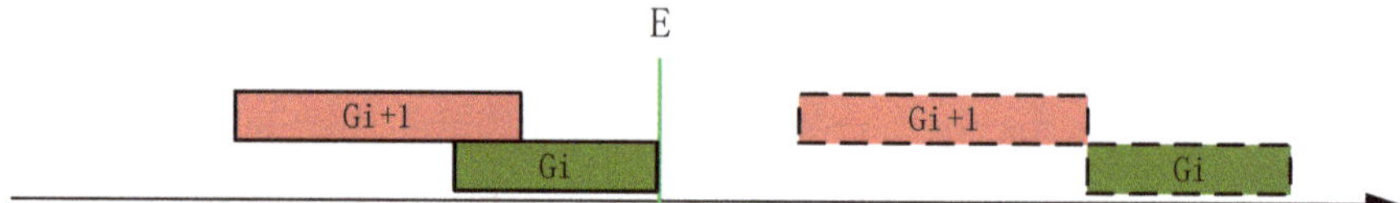

Figure 5. The two groups arrived at the intersection successively, causing congestion.

In conclusion, in order to minimize the total evacuation time, that is, to ensure the full use of the emergency exit, the group close to the emergency exit should give priority to escape. □

In the staged evacuation planning, the second key work is to calculate the delayed departure time of each evacuation group. Li et al. used the time extended network to calculate the delay time of each group [25]. The method first calculates the time window of each node on the evacuation path occupied by each escape group, and then calculates the latency of each group's departure time by the arrival sequence and the overlay of these time windows. The iterative process leads to redundant calculations in the algorithm, which results in its low efficiency. To avoid the problem, we comprehensively analyzed the operation of the staged evacuation process, and found Theorem 2 that simplifies the calculation process of the staged evacuation planning.

Theorem 2. *When evacuation groups are congested, the result of calculating their delayed departure times at the congested node is the same as that at the emergency exit.*

Proof of Theorem 2. According to Assumptions 3 and 4, in the evacuation network, the shortest paths from the exit to other nodes are equivalent to the shortest paths from other nodes to the exit. Therefore, the shortest paths of all groups can be obtained through Dijkstra algorithm to calculate the shortest paths from the exit to the nodes where each group is located. According to the operation principle of Dijkstra algorithm, the shortest path from the exit to each node will be obtained in the order of the shortest path length from small to large, so as to form the shortest path tree. Figure 6a is an indoor evacuation network. Ri represents a room and *E0* represents an exit. When Dijkstra algorithm is called, it will find in turn the shortest paths (*P1(E0-R6)*, *P2(E0-R8)*, *P3(E0-R6-R5)*, *P4(E0-R6-R2)*, *P5(E0-R3)*, *P6(E0-R6-R5-R1)*, *P7(E0-R8-R7)*, *P8(E0-R6-R5-R1-R4)*) from *E0* to *R6*, *R8*, *R5*, *R2*, *R3*, *R1*, *R7* and *R4* respectively. The route length of these paths will increase in turn, which are 6, 9, 13, 14, 15, 16, 21, 26. These paths compose a shortest path tree with *E0* as the root node. The tree is shown in Figure 6b, where the number marked on each node represents the order of obtaining its shortest path when running Dijkstra algorithm. When the shortest path from *E0* to each room node is calculated, the shortest path from each room to *E0* can be obtained by flipping the path direction. Figure 6c illustrates the two shortest paths. One is the path from R1 to the emergency exit *E0* and the other is that from *R2* to *E0*. They meet each other at node *R6* and overlap from *R6* to *E0*. □

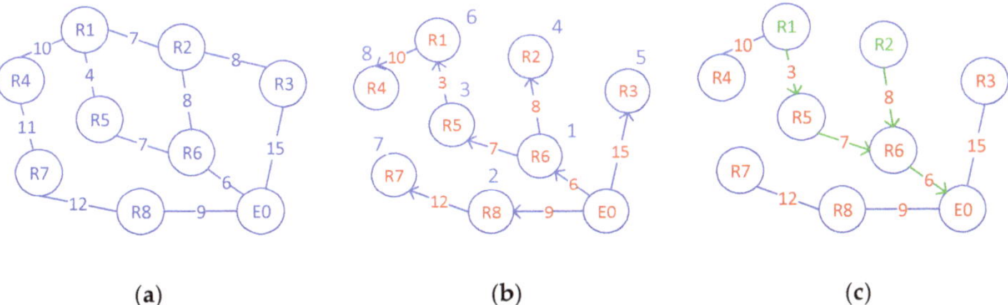

Figure 6. Illustration of generation of the shortest path tree and convergence of two escape groups. (**a**) Indoor evacuation network; (**b**) Shortest path tree; (**c**) Two shortest paths with intersection.

We can see from the execution process of Dijkstra algorithm that the shortest path of any node whose shortest path isn't determined will be obtained by the determined shortest paths. Therefore, once the evacuation paths of any two evacuation groups has an intersection, the two paths will be completely overlapped from the intersection to the exit, as shown in Figure 6c. In the process of evacuation, because the two groups have the same escape speed, their travel time after passing the intersection (that is, from the intersection to the emergency exit) is also equal. As a result, it is equivalent to calculate the delay time at the exit or intersection when the two groups are congested at the intersection. Assume that the evacuation speed $V = 1$, *G1* represents the evacuation group starting from *R1*, $t_e^1 = 3$; *G2* represents the evacuation group starting from *R2*, $t_e^2 = 5$; t_{pi}^1 is the evacuation time of G1 from R1 to R6; t_{pi}^2 is the evacuation time of G2 from R2 to R6; t_{pe}^1 is the evacuation time of G1 from R1 to E0; t_{pe}^2 is the evacuation time of G2 from R2 to E0. Then, the delay time of *G1* at the intersection $ti = t_{pi}^2 + t_e^2 - t_{pi}^1$, and the delay time of *G1* at the exit $te = t_{pe}^2 + t_e^2 - t_{pe}^1$. $t_{pe}^1 - t_{pi}^1 = t_{pe}^2 - t_{pi}^2$, so $ti = te$.

4.2. Proposed Algorithm for Multi-Exit Network

Based on the discussion above, we propose a partitioned and staged evacuation planning (PSEP) algorithm applied to multi-exit buildings. It should be noted about the algorithm that: a) evacuation planning is processed in groups to reduce the complexity of processing; b) for all groups in each evacuation zone, the staged evacuation is implemented; c) in order to minimize the total evacuation time, any two congested groups must successively pass through the emergency exit when delaying their departure time in each zone.

4.2.1. Algorithm Description

The whole algorithm is divided into two procedures. The first procedure will allocate an optimal exit to each evacuation group (all evacuation groups passing through the same exit belong to the same evacuation zone), and the second will compute the departure time of each evacuation group in the same zone. Their pseudo code is as follows.

Input: indoor road network model, exits, escape speed (V), group size (uniform or random) and the number of groups (n).

Output: Total evacuation time (T), the departure time (t_d^i) and the evacuation path for each group.

Notes about Procedure 1: Line 5 adds all evacuation groups in the network into an array N. Line 6 adds all exits in the network into an array E. Line 7 initializes the number of evacuees evacuated by each exit to 0. While N is not empty, namely there are any evacuation groups in N that aren't allocated to any exits, Lines 9 to14 compare the number of evacuees passing through each exit to find the exit $minE$ with the fewest evacuees. Lines 15 to 16 take $minE$ as the starting point, run Dijkstra algorithm to expand a new shortest path. The group located at the end node of the new path is $minG$. Then let $minG$ evacuate through $minE$. Next, update the number of evacuees passing through $minE$ (Line 19) and remove $minG$ from N (Line 20). Thus, an evacuation group is allocated to one exit by one loop. When all evacuation groups are allocated to an exit (that is, until N is empty), the loop ends.

Procedure 1 (Allocate an optimal exit to each evacuation group):

1	**Integer** *ne* //*ne* represents the number of exits in the network.
2	**Array** *N[n]* //*N* is used to store all evacuation groups.
3	**Array** *E[ne]* //*E* is used to store all safety exits.
4	**Array** *G[ne]* //*G* is used to store the current number of evacuees at each exit.
5	**Add** all evacuation groups **into** *N*
6	**Add** all evacuation exits **into** *E*
7	**Initialize** all elements of *G* to 0
8	**While** *N* is not empty **do**
9	**Let** *minE* = 1 //*minE* is a variable used to record the index of the exit with the minimum evacuees.
10	**For** *e* = 2 **to** *ne* //*e* is a local loop variable
11	**If** *G[e]* < *G[minE]*, **then**
12	**Let** *minE* = *e*
13	**End if**
14	**End for**
15	**Take** *minE* **as** the starting point
16	**Run** Dijkstra algorithm to expand a new shortest path//referring to Figure 6a
17	**Let** *minG* = *N[i]* //*N[i]* is the group located at the end node of the new path
18	**Let** *minG* evacuate passing through *minE*
19	**Update** the number of evacuees passing through *minE*
20	**Remove** *minG* **from** *N*
21	**End while**

In brief, during the implementation of Procedure 1, there is a Dijkstra algorithm at each exit, but only the Dijkstra algorithm on the exit with the least number of evacuees runs at each time, and the Dijkstra algorithm only expands one shortest path at a time (that is, to find an evacuation group). Then the number of evacuees allocated to each evacuation exit is compared with each other to determine which exit to run Dijkstra algorithm until all evacuation groups are allocated.

Notes about Procedure 2: Lines 3 adds all evacuation groups in the network into an array N with the array length of n. Line 4 sorts all evacuation groups in N according to their shortest route length. Then an outer loop (Line 5) is used to process each zone, namely, each exit. There are two inner loops in the outer loop. The first inner loop (Lines 6 to 10) is used to extract all each evacuation groups passing through the same exit from N, then adds them into M by order. Line 11 makes group 1 in M depart immediately once an emergency occurs, and $a = 1$, where a is the evacuation sequence number of the first group in each evacuation combination that successively passes through the emergency exit. Then the second inner loop (Lines 12 to 17) is executed to compute the departure time of *Gi* in *M*, where *i* is from 2 to *m* that is the number of groups in the same zone. The departure time of *Gi* is calculated as follows: $t_d^i = t_p^a + \left(t_e^a + \ldots + t_e^{i-1}\right) - t_p^i$. If $t_d^i > 0$, the delayed time of *Gi* is t_d^i. Otherwise, *Gi* evacuates immediately without any delay once an emergency occurs, and let $a = i$. Execute the inner loop until all evacuation groups in *M* get their departure time, then return to Line 5. When the outer loop finishes and the procedure ends.

Procedure 2 (Calculate the departure time of each group in each zone):

1	**Array** *N[n]* //*N* is used to store all evacuation groups
2	**Array** *M[n]* //*M* is used to record the groups assigned to the same exit
3	**Add** all evacuation groups **into** *N*
4	**Sort** *N* **By** their route length
5	**For** *e* = 1 **to** *ne* //to process each zone by the loop
6	**For** *j* = 1 **to** *n* //find all groups that passes through Exit *E[e]* by the loop
7	**If** *N[j]* passes through *E[e]* **then**
8	**Add** *N[j]* into *M*
9	**End if**
10	**End for**
11	**Let** $t_d^1 = 0, a = 1$
12	**For** *i* = 2 **to** *m* //to compute the departure time of all groups passing through Exit *E[e]* by the loop
13	$t_d^i = t_p^a + \left(t_e^a + \quad + t_e^{i-1}\right) - t_p^i$
14	**If** $t_d^i < 0$, **then**
15	$t_d^i = 0, a = i$
16	**End if**
17	**End for**
18	**End for**

4.2.2. Time Complexity Analysis

In Procedure 1, the while loop runs n times, and the for loop runs ne times. In addition, when running Dijkstra algorithm to expand each node, the path length of all nodes in the network needs to compare n times. So, the time complexity of Procedure 1 is $O(n(n + ne))$. For $ne \ll n$, the final time complexity is $O(n^2)$.

In Procedure 2, the outer for loop runs *ne* times, the first inner for loop runs *n* times, the second for loop runs *m* times. So, the time complexity of Procedure 2 is $O(ne(n + m))$. For $m \leq n$, the final time complexity is $O(n * ne)$. Of course, this process also includes a sorting process with the time complexity $O(n^2)$ or $O(n \log_2 n)$.

Once the PSEP algorithm completes partition, each zone is equivalent to a single-exit evacuation network. For a zone, the time complexity of calculating the departure time of each evacuation group is $O(m)$, while the time complexity of completing the calculation by the algorithm in [25] is $O(m^2 \bar{k})$, where *m* is the number of groups in the zone, *k* is the number of arcs of all evacuation path and $\bar{k}$ is the arithmetic mean of *k*.

5. Case Study

To verify the validity and efficiency of the PSEP algorithm, we conducted two tests. One was used to verify the correctness and efficiency of the algorithm for the single-exit evacuation; the other to discuss the rationality of the partition method of the PSEP algorithm and to compare its performance with an existing algorithm. Since the PSEP algorithm is based on the single-exit evacuation algorithm, both tests are valuable to illustrate the advantages of the PSEP algorithm. Test data is the three-dimensional path network of the teaching building J6 of Shandong University of Science and Technology (SDUST), as shown in Figure 7, where each vertex (i.e., node) in the network represents an escape group and each edge (i.e., arc) represents a segment of indoor path. The network model consists of five layers, 818 nodes and 853 edges. On the first floor, there are three safety exits: *E1*, *E2* and *E3*). During the tests, nodes in the route network are randomly selected as the starting nodes in order to simulate the evacuation environment in reality.

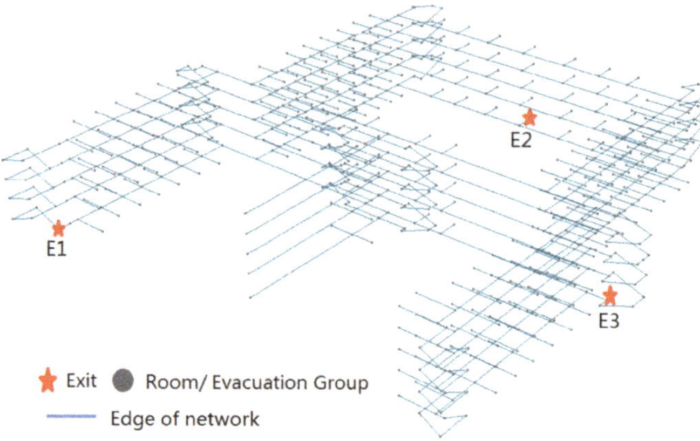

Figure 7. The indoor network model of the teaching building J6 of SDUST.

All involved algorithms were developed in C-Sharp and run on the portable notebook whose configuration were as follows: CPU i7-6500u, main frequency 2.5 GHz, running memory 12 G and the solid-state disk with capacity of 256 G.

5.1. Tests Based on Single-Exit Network

The PSEP algorithm is mainly inspired by the algorithm of [25]. We firstly compare and discuss their efficiency and results. Additionally, the algorithm of [25] is only suitable for the single-exit network, so we choose a part of the teaching building J6 (i.e., an evacuation zone with one safety exit) as the test zone to test the influence of the number of escape groups, the size of groups, the escape speed and other factors on the total evacuation time, as well as the efficiency of the two algorithms. The test network includes only one exit *E1*, 210 edges and 210 vertices. When the PSEP algorithm is applied to the single-exit network, the first procedure of the algorithm is omitted for the network has only one exit.

5.1.1. Influence of the Number of Groups on Total Evacuation Time

The size of all groups is 15 m, the escape speed is fixed as 3 m/s, and the number of evacuation groups is set to 90, 130, 170, 210 respectively. The test results are shown in Figure 8.

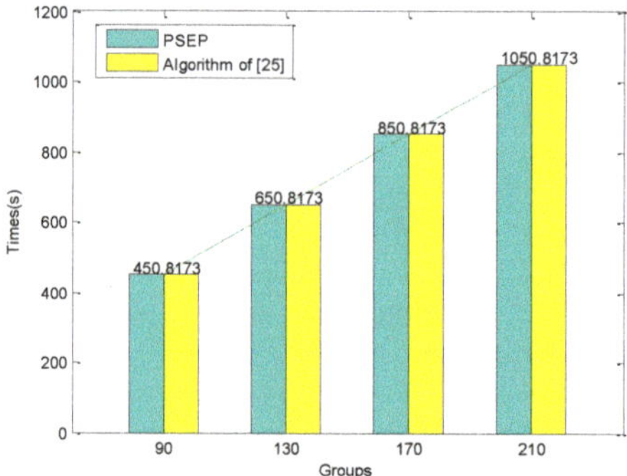

Figure 8. Comparison of their total evacuation time varying with the number of evacuation groups.

Figure 8 shows that when the group size and the escape speed of each evacuation group are fixed, their total evacuation times increase linearly with the number of evacuation groups for both algorithms. At the same time, the total evacuation times of the two algorithms are equal.

5.1.2. Influence of Evacuation Speed on Total Evacuation Time

The number of evacuation groups is 210, the group size is 15 m, and the escape speed is set to 2, 3, 4, and 5 m/s respectively. The test results are shown in Figure 9.

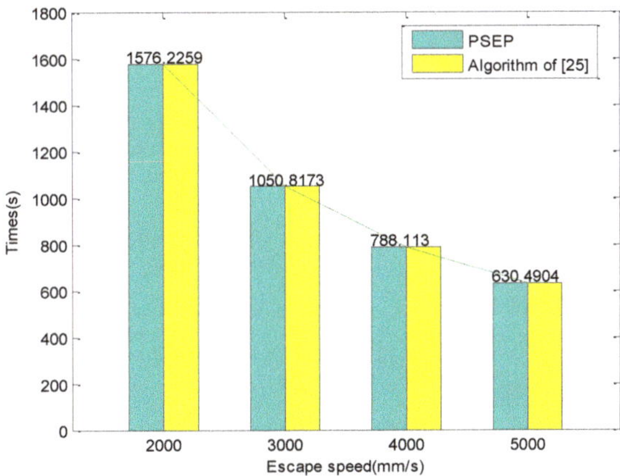

Figure 9. Comparison of their total evacuation time varying with the escape speed.

Figure 9 illustrates that when the number of evacuation groups and the group size are fixed, the total evacuation times of the two algorithms decrease with the increase of the escape speed.

5.1.3. Influence of Group Size on Total Evacuation Time

The number of evacuation groups is 210, the escape speed is 3 m/s, and the group size is set to 6, 15, 24, and 33 m. The test results are shown in Figure 10.

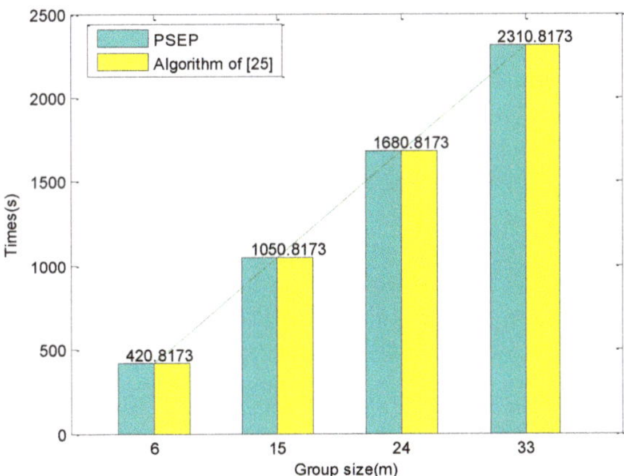

Figure 10. Comparison of their total escape time varying with the group size.

Figure 10 illustrates that when the number of evacuation groups and the escape speed are fixed, their total evacuation times increase linearly with the increase of the group size for both algorithms.

5.1.4. Comparison of Operating Efficiency

The group size is 15 m, the escape speed is 3 m/s, and the number of evacuation groups is set to 120, 280, 440, 600, 760, 920 and 1080 respectively. The test statistics are shown in Table 1. *Td* represents the time consumed by the algorithm of [25] and *Tp* represents that by the PSEP algorithm. Figure 11a shows the curves of the time consumed by the two algorithms, from which we can see that their time consumption is increasing with the increase of the number of evacuation groups, but the time consumed by the PSEP algorithm is significantly less than that by the algorithm of [25]. Figure 11b indicates the ratio curve of the time consumed by the two algorithms, from which we can see that the more evacuation groups there are, the more obvious the efficiency advantage of the PSEP algorithm over the algorithm of [25] is. When the number of groups is 1080, *Td/Tp* reaches 41465.

Table 1. Statistical results of time consumed by the two algorithms.

Group Number	120	280	440	600	760	920	1080
Tp (ms):	1	2	7	12	18	27	36
Td (ms):	329	7711	42,366	142,399	363,591	773,820	1,492,754
Td/Tp:	329	3856	6052	11,867	20,200	28,660	41,465

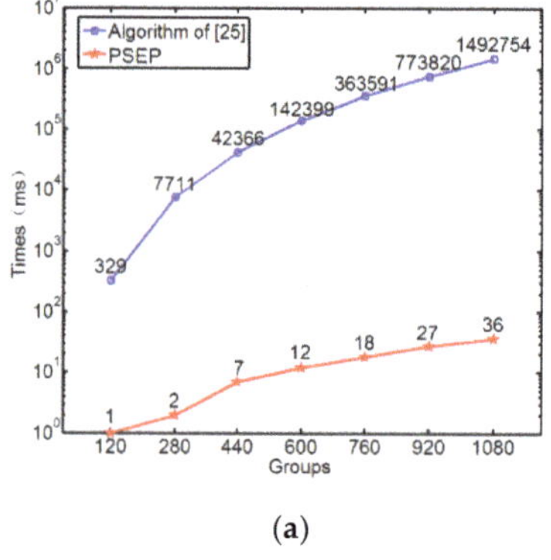

(a)

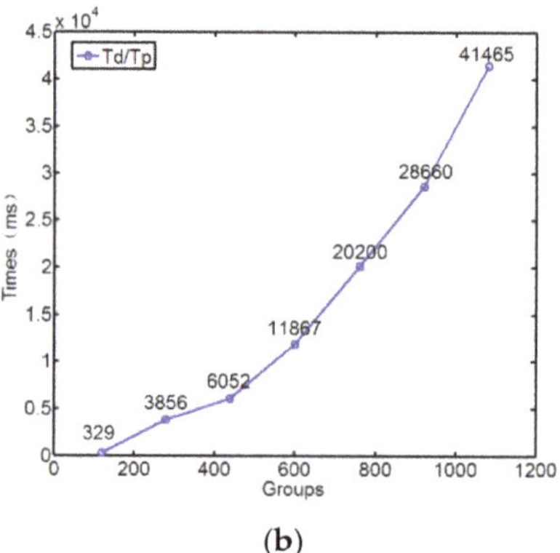

(b)

Figure 11. Comparison of the efficiency of the two algorithms. (**a**) The curves of the consumed time by the two algorithms; (**b**) The efficiency ratio of the two algorithms

It can be seen from the first three tests that the total evacuation time of the two algorithms is the same regardless of the evacuation condition, which proves the correctness of the proposed algorithm. Test 4 illustrates that the PSEP algorithm is much more efficient than the algorithm of [25]. The reason is that there are a lot of repeated calculations in the algorithm of [25]. Every time the departure time of an evacuation group is determined, it is necessary for each evacuation group whose departure time has not been determined to recalculate its time windows at all nodes on its evacuation path, while the PSEP algorithm only needs to calculate the time window of each group occupying the exit to determine the departure time of all escape groups. In addition, the tests above also prove the correctness of Theorems 1 and 2.

5.2. Tests Based on Multi-Exit Network

According to the principle of the PSEP algorithm, the partition method and the density of indoor evacuees are two important factors that affect the overall evacuation time. Therefore, we tested their influence on evacuation efficiency. Furthermore, the relation between the evacuation path length and delayed time of each evacuation group is tested. Its performance is also compared with that of an existing algorithm.

5.2.1. Influence of Partitioning Methods on Evacuation Efficiency

In theory, the partition method based on the principle of "nearest evacuation" will increase the overall evacuation time when the density of indoor evacuees is larger. In order to verify the theory, we apply the principles of "nearest evacuation" and "balanced evacuation" to the PSEP algorithm respectively to test their influence on evacuation. Figure 12a shows the partitioned indoor network only based on "nearest evacuation" that will find the nearest exit for each group, and Figure 12b shows that based on "balanced evacuation" that makes every exit have the approximately equal number of evacuees by considering both the number of evacuees and their path length. It can be seen from Figure 12 that the partition of some nodes in the route network has changed. Many nodes originally belonging to the zone E3 are assigned to E1, while the zone E1 also occupies part of the nodes in the original zone E2, and the zone E2 regains part of the nodes from the original zone E3. Furthermore, these adjusted nodes are mainly distributed in the adjacent area of the original zones.

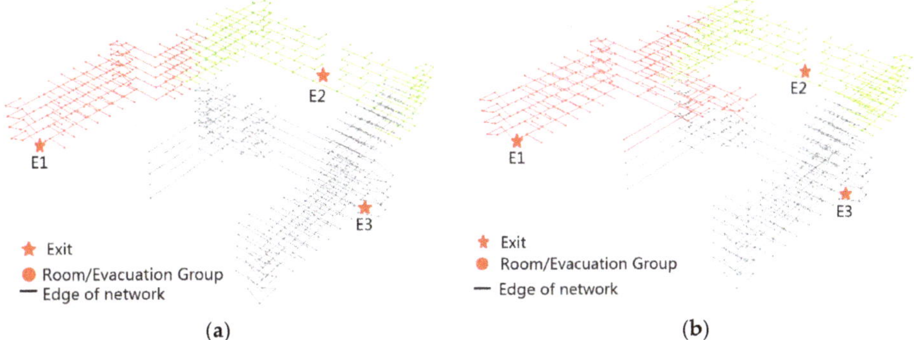

Figure 12. Comparison of results partitioned by two different partitioning methods. (**a**) The partitioned network based on nearest evacuation; (**b**) The partitioned network based on balanced evacuation

Table 2 shows the number of evacuation groups, the number of evacuees of each evacuation zone and their total evacuation time when the PSEP algorithm adopts the two partitioning methods respectively. It can be seen from the table that the number of evacuation groups of each exit when using the principle of "nearest evacuation" is not balanced while that are balanced when using the principle of "balanced evacuation". The nearest evacuation makes exits E1 and E2 are not fully utilized, which increases the overall evacuation time by 265 seconds compared with that of the balanced evacuation. Therefore, we can conclude that the partition strategy of evacuation zones will greatly affect the evacuation efficiency of the evacuation scheme.

Table 2. The number of evacuees, the number of evacuation groups in different zones and their total evacuation time based on balanced evacuation and nearest evacuation

	ID of Exits	E1	E2	E3
Balanced evacuation	The number of groups	264	277	277
	The number of evacuees	2352	2345	2393
	Evacuation time (s)	785.13	786.18	798.57
Nearest evacuation	The number of groups	210	236	372
	The number of evacuees	1845	2057	3188
	Evacuation time (s)	616.13	690.18	1063.57

5.2.2. Influence of Evacuation Density on Total Evacuation Time

Let the evacuation density $ER = EL/NL$, where EL is the total length of all evacuation groups and NL is the total length of all edges of the evacuation network. To test the influence of the evacuation density on the total evacuation time of different evacuation plans, we implemented the nearest evacuation and the balanced evacuation respectively with different evacuation densities. The total length of the evacuation route network is 5443.3 m, and the number of evacuation groups is 818. The length of all evacuation groups is set as 0.5, 1, 2, 3, 4 and 5 m respectively. The test results are shown in Table 3, where *Tn* represents the total evacuation time of nearest evacuation and *Tb* represents that of balanced evacuation. We can see that balanced evacuation has obvious advantages over nearest evacuation when the density of evacuees is large but nearest evacuation has a shorter overall evacuation time when the density of evacuees is small. Figure 13 shows that the greater the density of evacuees, the more advantageous balanced evacuation will be.

Table 3. Comparison of the total evacuation time of two strategies under different evacuation density

Group Length (m)	ER	*Tn* (s)	*Tb* (s)	*Tn-Tb* (s)
0.5	7.5%	70.703	85.101	−14.398
1	15%	128.64	102	26.64
2	30%	250.35	188.35	62
3	45%	373.49	280.49	93
4	60%	496.91	372.91	124
5	75%	620.91	465.91	155

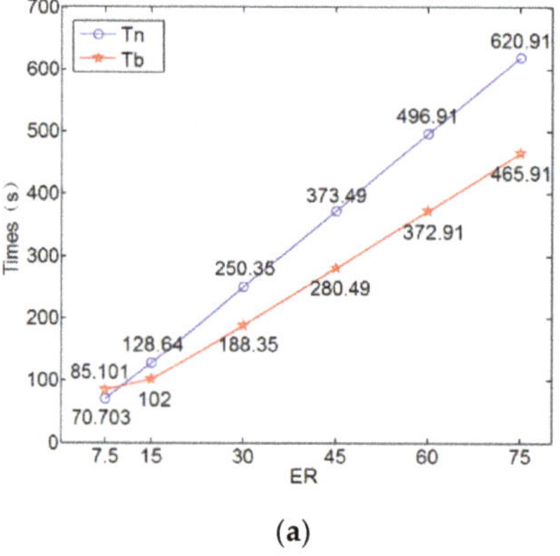
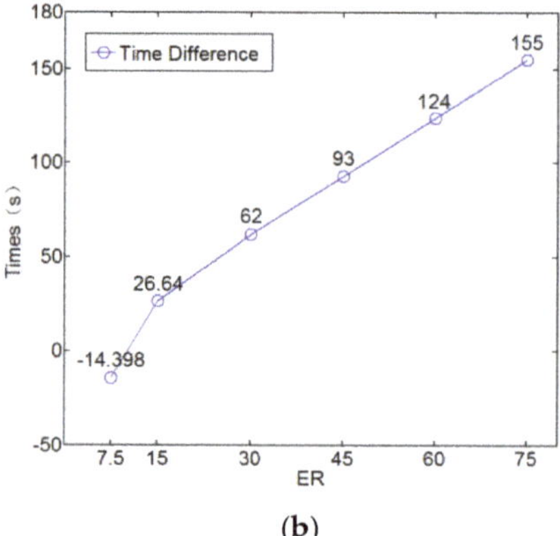

Figure 13. Comparison of the total evacuation time of two strategies under different evacuation density. (**a**) The change of their total evacuation time with the increase of evacuation density; (**b**) The change of the total evacuation time difference with the increase of evacuation density

5.2.3. Evacuation Process Simulation

In order to visually verify the effectiveness of our algorithm, we took the whole teaching building J6 of SDUST as the test scene to simulate the emergency evacuation process using our evacuation simulation software. In the simulation scene, color is used to identify different evacuation groups, and the length of line segment represents group size. Assuming that the start time of evacuation is t0, escape speed is 3 m/s, group size is random and the sum of evacuation groups is 818. In case of emergency, the visual simulation results of evacuation process of all groups starting to escape at the same time are shown in Figure 14, where (a)–(d) are the distribution of all escape groups at t0 + 8 s, t0 + 16 s, t0 + 24 s, and t0 + 32 s, respectively. From Figure 14, we can see that many colorful line segments are mixed together, which indicates that there is a large area of congestion. Obviously, serious congestion will reduce the escape speed of evacuees and ultimately lead to the extension of the total evacuation time.

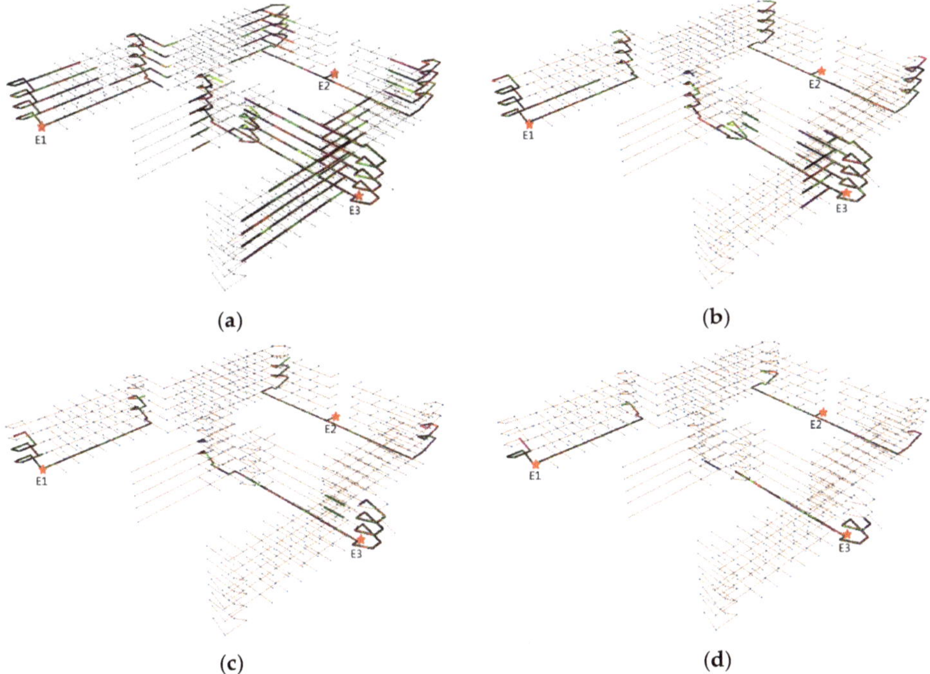

(a)   (b)

(c)   (d)

Figure 14. The simulation of simultaneous evacuation for the multi-exit network. Four screenshots are given at different times as follows: (**a**) t0 + 8 s; (**b**) t0 + 16 s; (**c**) t0 + 24 s; (**d**) t0 + 32 s.

Figure 15 shows the visual simulation results of the PSEP algorithm, where (a)–(f) are the distribution of all escape groups at t0 + 32 s, t0 + 64 s, t0 + 112 s, t0 + 144 s, t0 + 232 s, t0 + 272 s, respectively. The simulation process shows that all groups escape orderly according to the assigned departure time without any congestion on the way and all groups pass through the emergency exit successively. All of these ensure the efficient operation of the whole evacuation process and reduce the overall evacuation time.

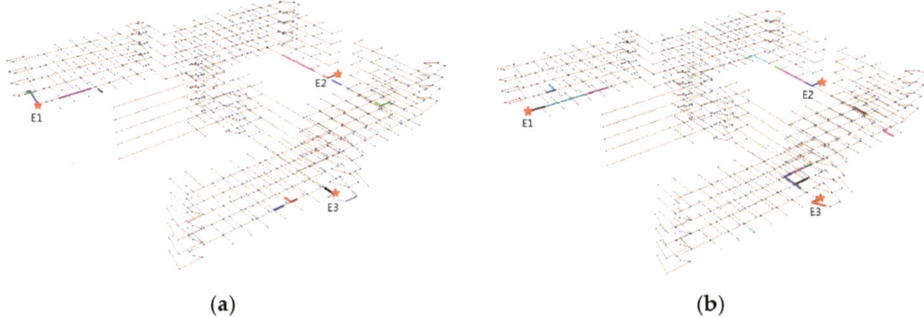

(a)   (b)

Figure 15. *Cont.*

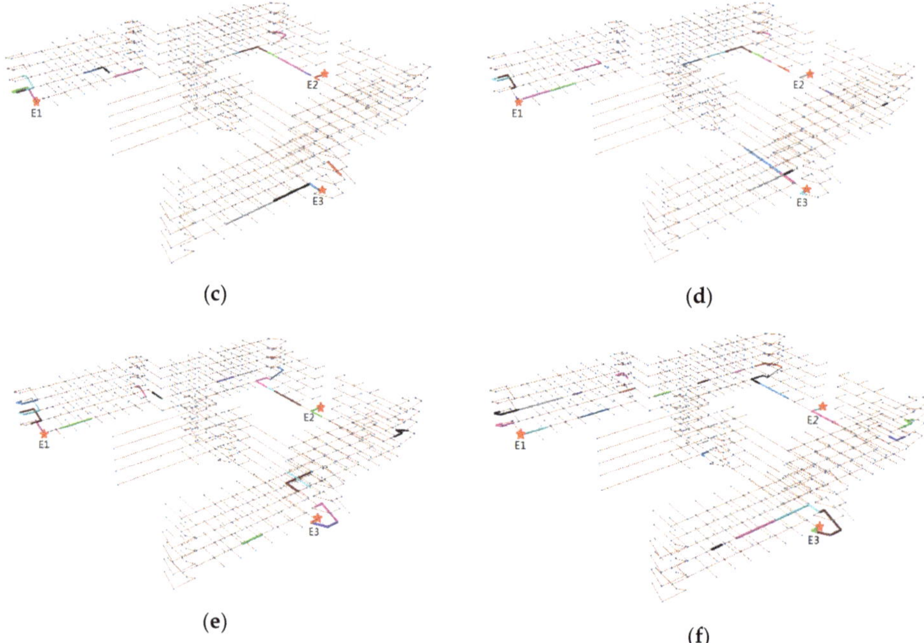

Figure 15. The simulation of partitioned and staged evacuation for the multi-exit network. Six screenshots are given at different times as follows: (**a**) t0 + 32 s; (**b**) t0 + 64 s; (**c**) t0 + 112 s; (**d**) t0 + 144 s; (**e**) t0 + 232 s; (**f**) t0 + 272 s.

The PSEP algorithm adopts the partitioned and staged evacuation strategy. Once evacuation partition is completed, the evacuation process in each zone is independent of each other. The total evacuation time of all escape groups is the maximum evacuation time of each zone. Table 4 shows the relationship between the evacuation path length and the delayed departure time of some evacuation groups in the zone E1, and Figure 16 shows that of all evacuation groups in zones E1, E2, and E3 respectively. We can see that the delayed departure time of all groups in each zone increases with the increase of the evacuation path length.

Table 4. The relationship of the evacuation path length and the delayed departure time of some groups planned by the PSEP algorithm when evacuation density is large.

Group ID	Group Size (m)	Escape Speed(m/s)	Path Length (m)	Delayed Time (s)
1	11	3	3.40	0.00
2	8	3	5.80	2.87
3	8	3	7.61	4.93
4	5	3	13.05	5.78
5	11	3	14.46	6.98
6	5	3	14.57	10.61
7	7	3	19.27	10.71
8	13	3	19.33	13.02
9	4	3	19.80	17.20
10	12	3	22.07	17.78
11	7	3	26.02	20.46
12	9	3	26.04	22.79
... ...	... ...	... ...	... ...	... ...

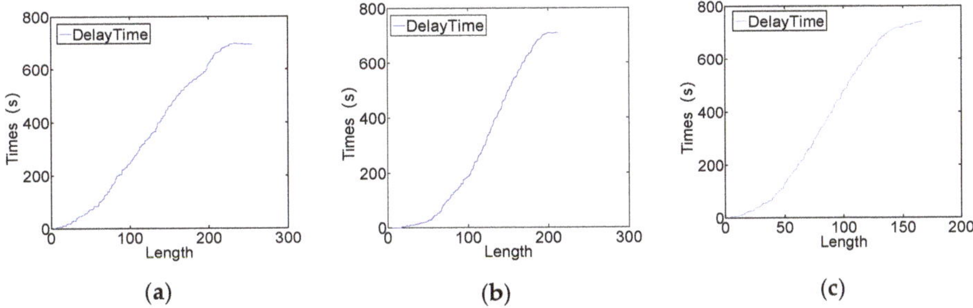

Figure 16. The relationship of the evacuation path length and the delayed departure time of all groups in each zone when evacuation density is large. (**a**) Zone E1; (**b**) Zone E2; (**c**) Zone E3.

5.2.4. Relation between Evacuation Path Length and Delayed Time

Table 4 shows the relationship between the length of the evacuation path and the delayed time in the case of a large density of evacuees. To get a more comprehensive picture of their relationship, we conducted the other test in the case of a low density of evacuees. Let all evacuation groups be 2 m in size and the other test conditions will remain the same. Table 5 shows the evacuation path length and the delayed departure time of the partial evacuation groups in zone E1. Figure 17 shows the relationship between the evacuation path length and the delayed departure time for all evacuation groups in zones E1, E2, and E3, respectively.

Table 5. The relationship of the evacuation path length and the delayed departure time of some groups planned by the PSEP algorithm when evacuation density is low

Group ID	Group Size (m)	Escape Speed (m/s)	Path Length (m)	Delayed Time (s)
1	2	3	3.40	0.00
2	2	3	5.80	0.00
3	2	3	7.61	0.06
4	2	3	13.05	0.00
5	2	3	14.46	0.20
6	2	3	14.57	0.83
7	2	3	19.27	0.00
8	2	3	19.33	0.65
9	2	3	19.80	1.16
10	2	3	22.07	1.07
11	2	3	26.02	0.42
12	2	3	26.04	1.08
… …	… …	… …	… …	… …

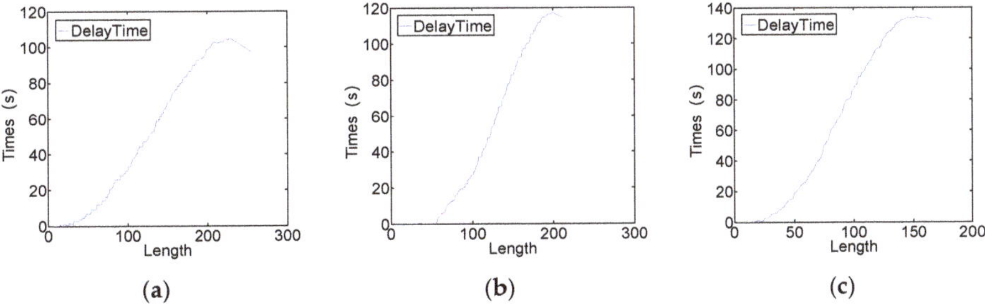

Figure 17. The relationship of the evacuation path length and the delayed departure time of all groups in each zone when evacuation density is low. (**a**) Zone E1; (**b**) Zone E2; (**c**) Zone E3.

Table 5 shows that the delayed time of the evacuation groups will not increase completely with the increase of their evacuation path length, and some of them will be exceptional. This exception occurs because of the time interval between some of the adjacent evacuation groups (Figure 3), which will reduce the delayed departure time of the group that arrives later. For example, if the time windows of two groups A and B occupying the same exit are [23.0, 25.0] and [29.0, 32.0] respectively when they start to escape at the same time, they will not congest with each other. But, if the groups ahead of A makes it have to delay 5 s to depart in order to avoid congestion at the exit, the time window of the group A occupying the exit becomes [28.0, 30.0]. To avoid congestion with the group A, and the group B should be delayed by 1 second that is less than the delayed time of the group A.

5.2.5. Performance Comparison

Li et al. extended their approach suitable for the single-exit evacuation to the multi-exit evacuation in [26]. Here, our algorithm is compared with that in [26] based on a testing network model consisting of 923 nodes and 1779 edges. Three of these nodes are exits. When the group size is uniform, the test results are shown in Table 6, where Ng and Ne are the number of groups and the number of evacuees passing through each exit and Te is the evacuation time at each exit. Figure 18 shows the change of the total evacuation time and the operation time of each algorithm when the group size increases gradually.

Table 6. The test results when the group size is uniform.

ER	Group Size	Algorithm	Parameters	E1	E2	E3	Clearing Time (s)
32.6%	6	PSEP	Ng	305	307	308	620.88
			Ne	1830	1842	1848	
			Te (s)	613.20	619.06	620.88	
		Algorithm of [26]	Ng	307	306	307	618.88
			Ne	1842	1836	1842	
			Te (s)	617.20	618.77	618.88	
48.9%	9	PSEP	Ng	305	307	308	927.88
			Ne	2745	2763	2772	
			Te (s)	917.97	925.06	927.88	
		Algorithm of [26]	Ng	304	305	311	936.88
			Ne	2736	2745	2799	
			Te (s)	922.05	919.06	936.88	
65.2%	12	PSEP	Ng	305	307	308	1235.60
			Ne	3660	3684	3696	
			Te (s)	1222.97	1231.98	1235.60	
		Algorithm of [26]	Ng	309	307	304	1238.97
			Ne	3708	3684	3648	
			Te (s)	1238.97	1233.15	1227.59	
81.5%	15	PSEP	Ng	305	307	308	1543.60
			Ne	4575	4605	4620	
			Te (s)	1527.97	1538.98	1543.60	
		Algorithm of [26]	Ng	307	306	307	1538.60
			Ne	4605	4590	4605	
			Te (s)	1537.97	1533.97	1538.60	
97.8%	18	PSEP	Ng	305	307	308	1851.60
			Ne	5490	5526	5544	
			Te (s)	1832.97	1845.98	1851.60	
		Algorithm of [26]	Ng	307	306	307	1850.28
			Ne	5526	5508	5526	
			Te (s)	1844.97	1850.28	1845.60	

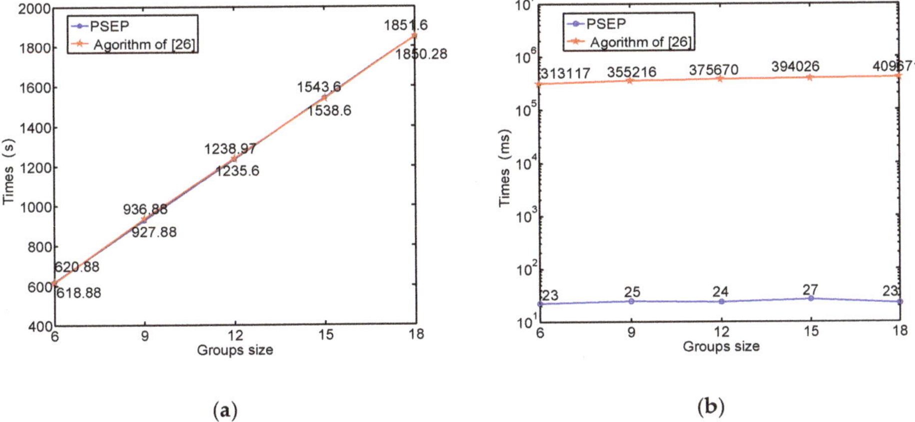

(a)   **(b)**

Figure 18. The comparative results when the group size is uniform. (**a**) The change of the total evacuation time with the increasement of the number of groups; (**b**) The change of the operation time with the increasement of the number of groups.

To make the test more realistic, we let the group size go to random numbers. When the group size is random, the test results are shown in Table 7, where the value of the group size is a range, which means that the group size can take any value within this range randomly. Figure 19 shows the change of total evacuation time and the operation time of each algorithm when the group size increases gradually.

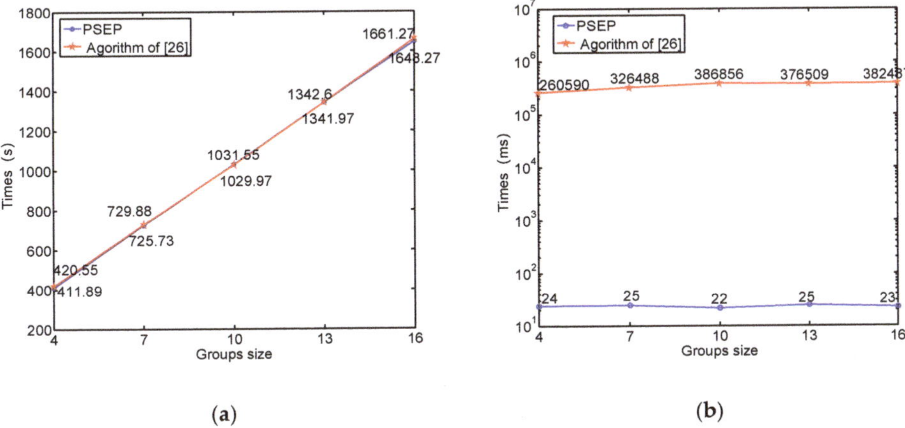

(a)   **(b)**

Figure 19. The comparative results when the group size is random. (**a**) The change of the total evacuation time with the increasement of the number of groups; (**b**) The change of the operation time with the increasement of the number of groups.

As shown in Tables 6 and 7 and Figures 18 and 19, our algorithm and that of [26] are very close in the overall evacuation time, but the planning efficiency of our algorithm is much higher than that of [26]. For applications that require rapid or real-time evacuation planning, our algorithm has obvious advantages.

Table 7. The test results when the group size is random.

ER	Group Size	Algorithm	Parameters	E1	E2	E3	Clearing Time (s)
21.5%	3–6	PSEP	Ng	313	306	301	411.89
			Ne	1211	1218	1219	
			Te (s)	408.20	411.89	411.89	
		Algorithm of [26]	Ng	311	299	310	420.55
			Ne	1214	1189	1245	
			Te (s)	411.28	406.19	420.55	
38.2%	6–9	PSEP	Ng	304	309	307	725.73
			Ne	2151	2162	2157	
			Te (s)	719.97	725.73	723.55	
		Algorithm of [26]	Ng	304	304	312	729.88
			Ne	2137	2157	2176	
			Te (s)	715.30	728.88	729.88	
54.3%	9–12	PSEP	Ng	305	309	306	1031.55
			Ne	3036	3074	3083	
			Te (s)	1014.97	1028.64	1031.55	
		Algorithm of [26]	Ng	309	303	308	1029.97
			Ne	3081	3050	3062	
			Te (s)	1029.97	1020.64	1024.55	
70.8%	12–15	PSEP	Ng	305	306	309	1342.60
			Ne	3973	3999	4017	
			Te (s)	1327.30	1336.98	1342.60	
		Algorithm of [26]	Ng	308	306	306	1341.97
			Ne	4017	3996	3976	
			Te (s)	1341.97	1335.98	1328.94	
87.1%	15–18	PSEP	Ng	304	307	309	1648.27
			Ne	4881	4926	4934	
			Te (s)	1629.97	1645.98	1648.27	
		Algorithm of [26]	Ng	305	304	311	1661.27
			Ne	4899	4869	4973	
			Te (s)	1635.97	1640.37	1661.27	

6. Conclusions

For indoor emergency evacuation with a large number of evacuees, a partitioned and staged evacuation planning algorithm considering indoor congestion is proposed. According to the idea of "balanced evacuation", the algorithm coordinates the number of evacuees at different exits by the improved Dijkstra algorithm, which partitions the whole evacuation area and turns the multi-exit evacuation into the single-exit evacuation, thus simplifying the complexity of problem processing. For the single-exit evacuation, the proposed algorithm only needs to consider the time conflict between the time windows of all evacuation groups at exits, then it can calculate the departure time of each group. Compared with the traditional algorithm that considers the conflict between the time windows of all evacuation groups at every node of the evacuation paths to calculate the departure time of each group, it reduces the calculation at redundant path nodes and greatly improves the efficiency of emergency evacuation planning. In practice, the PSEP algorithm in this paper provides not only the best evacuation path but also the optimal departure time for each group to ensure that all groups will not be congested during evacuation, which has a strong operability. The smart city makes it possible to access indoor evacuation information in real time such as the distribution of evacuees and the development of an indoor disaster, which provides a data base for the real-time design of an evacuation scheme. The design requires high efficiency of planning algorithms. Our algorithm is simple and has great advantages in operating efficiency, which will meet the development and demand for intelligent emergency evacuation systems and emergency command.

Although the PSEP algorithm achieves better results in the case of crowded indoor occupants by transforming the multi-exit indoor evacuation problem into the single-exit indoor evacuation problem based on the "balanced evacuation" principle, it is still an approximate optimal result due to the lack of rigorous mathematical reasoning and proofs. Meanwhile, the partition strategy may not obtain the global optimal solution when the indoor occupants are sparse. Therefore, we will consider the influence of the density and distribution of evacuees on the total evacuation time and the connection among all exits in the future to optimize the total evacuation time further. In addition, when an emergency occurs, let the groups that may be congested wait in the original place, which is not applicable to the occurrence of local disasters such as indoor fire. It should be considered to set up an indoor disaster risk area, evacuate the evacuees in the risk area to the safety area first, and then evacuate them to the safety exit. We will attempt to resolve this in a further study.

Author Contributions: Conceptualization, Litao Han and Haisi Zhang; Methodology, Litao Han, Huan Guo and Qiaoli Kong; Software, Huan Guo, Haisi Zhang and Cheng Gong; Validation, Huan Guo, Haisi Zhang and Cheng Gong; Writing–original draft, Litao Han, Huan Guo and Aiguo Zhang; Writing–review & editing, Litao Han, Qiaoli Kong and Aiguo Zhang. All authors have read and agreed to the published version of the manuscript.

Funding: This research was funded by the Natural Science Foundation of Shandong Province (Grant Nos. ZR2017MD003 and ZR2017MD032), the National Natural Science Foundation of China (Grant No. 41704015), and the Natural Science Foundation of Fujian Province (Grant No. 2016J01198).

Acknowledgments: We are grateful for the assistances of the reviewers and editors, and especially would like to express our gratitude to Xiang Li and his team from Key Lab of Geographical Information Science, Ministry of Education, East China Normal University for providing their source code that makes us able to compare our algorithm.

References

1. Kinateder, M. Human behaviour in severe tunnel accidents: Effects of information and behavioural training. *Transp. Res. Part F Traffic Psychol. Behav.* **2013**, *17*, 20–32. [CrossRef]
2. Li, X.; Li, Q.; Xu, X.; Xu, D.; Zhang, X. A novel approach to developing organized multispeed evacuation plans. *Trans. GIS* **2018**, *22*, 1205–1220. [CrossRef]
3. Shin, Y.; Kim, S.; Moon, I. Simultaneous evacuation and entrance planning in complex building based on dynamic network flows. *Appl. Math. Model.* **2019**, *73*, 545–562. [CrossRef]
4. Bi, H. Evacuee flow optimisation using G-network with multiple classes of positive customers. In Proceedings of the IEEE 24th International Symposium on Modeling, Analysis and Simulation of Computer and Telecommunication Systems, London, UK, 19–21 September 2016; pp. 135–143. [CrossRef]
5. Pelechano, N.; Malkawi, A. Evacuation simulation models: Challenges in modeling high rise building evacuation with cellular automata approaches. *Autom. Constr.* **2008**, *17*, 377–385. [CrossRef]
6. Wang, R.; Zhou, L.; Liu, J. Study on cellular automation evacuation model based on improved ant colony optimization algorithm. *China Saf. Sci. J.* **2018**, *28*, 38–43. [CrossRef]
7. Pan, X.; Han, C.; Dauber, K.; Law, K.H. A multi-agent based framework for the simulation of human and social behaviors during emergency evacuations. *Ai Soc.* **2007**, *22*, 113–132. [CrossRef]
8. Ren, C.; Yang, C.; Jin, S. Agent-based modeling and simulation on emergency evacuation. *Complex Sci.* **2009**, *5*, 1451–1461. [CrossRef]
9. Chen, X.; Zhan, F. Agent-based modeling and simulation of urban evacuation: Relative effectiveness of simultaneous and staged evacuation strategies. *J. Oper. Res. Soc.* **2008**, *59*, 25–33. [CrossRef]
10. Parisi, D.R.; Dorso, C.O. Microscopic dynamics of pedestrian evacuation. *Phys. A Stat. Mech. Appl.* **2005**, *354*, 606–618. [CrossRef]
11. Yang, X.; Dong, H.; Wang, Q.; Chen, Y.; Hu, X. Guided crowd dynamics via modified social force model. *Phys. A Stat. Mech. Appl.* **2014**, *411*, 63–73. [CrossRef]
12. Helbing, D.; Isobe, M.; Nagatani, T.; Takimoto, K. Lattice gas simulation of experimentally studied evacuation dynamics. *Phys. Rev. E* **2003**, *67*, 067101. [CrossRef]

13. Guo, X.; Chen, J.; Zheng, Y.; Wei, J. A heterogeneous lattice gas model for simulating pedestrian evacuation. *Phys. A Stat. Mech. Appl.* **2012**, *391*, 582–592. [CrossRef]

14. Li, W.; Li, Y.; Yu, P.; Gong, J.H.; Shen, S.; Huang, L.; Liang, J.M. Modeling, simulation and analysis of the evacuation process on stairs in a multi-floor classroom building of a primary school. *Phys. A Stat. Mech. Appl.* **2017**, *469*, 157–172. [CrossRef]

15. Liu, S.; Yang, L.; Fang, T.; Li, J. Evacuation from a classroom considering the occupant density around exits. *Phys. A Stat. Mech. Appl.* **2009**, *388*, 1921–1928. [CrossRef]

16. Fleischer, L.; Skutella, M. Minimum cost flows over time without intermediate storage. In Proceedings of the 14th Annual ACM-SIAM Symposium on Discrete Algorithms, Baltimore, MD, USA, 12–14 January 2003; Society for Industrial and Applied Mathematics Philadelphia: Philadelphia, PA, USA, 2003; Volume 8, pp. 66–75.

17. Zhang, X.; Chang, G. A dynamic evacuation model for pedestrian–vehicle mixed-flow networks. *Transp. Res. Part C Emerg. Technol.* **2014**, *40*, 75–92. [CrossRef]

18. Zheng, X.; Cai, L.; Zhang, M.; Jing, H.L.; Chen, Y. Emergency evacuation path optimization model under multi-export conditions. *China Saf. Sci. J.* **2019**, *29*, 180–186. [CrossRef]

19. Stepanov, A.; Smith, J.M. Multi-objective evacuation routing in transportation networks. *Eur. J. Oper. Res.* **2009**, *198*, 435–446. [CrossRef]

20. Fang, Z.; Li, Q.; Li, Q.; Han, L.; Shaw, S. A space–time efficiency model for optimizing intra-intersection vehicle–pedestrian evacuation movements. *Trans. Res. Part C Emerg. Technol.* **2013**, *31*, 112–130. [CrossRef]

21. Fang, Z.; Zong, X.; Li, Q.; Li, Q.; Xiong, S. Hierarchical multi-objective evacuation routing in stadium using ant colony optimization approach. *J. Trans. Geogr.* **2011**, *19*, 443–451. [CrossRef]

22. Li, Q.; Fang, Z.; Li, Q. Ant colony based evacuation route optimization model for mixed pedestrian-vehicle flows. In Proceedings of the Sixth International Conference on Pedestrian and Evacuation Dynamics, ETH, Zurich, Switzerland, 6–8 June 2012; Weidmann, U., Kirsch, U., Schreckenberg, M., Eds.; Springer: Cham, Switzerland, 2014; pp. 1213–1224. [CrossRef]

23. Wang, L.; Wang, R.; Tao, K. Research on capacity constrained evacuation route planning method in large-scale networks. *Sci. Surv. Mapp.* **2019**, *35*, 235–241. [CrossRef]

24. Sbayti, H.; Mahmassani, H.S. Optimal scheduling of evacuation operations. *Trans. Res. Rec.* **2006**, *1964*, 238–246. [CrossRef]

25. Li, X.; Huang, B.; Liu, Z.; Zhang, X.; Sun, J. A novel method for planning a staged evacuation. *J. Syst. Sci. Complex.* **2012**, *25*, 1093–1107. [CrossRef]

26. Li, X.; Li, Q.; Claramunt, C. A time-extended network model for staged evacuation planning. *Saf. Sci.* **2018**, *108*, 225–236. [CrossRef]

27. Chow, W.K.; Ng, C.M. Waiting time in emergency evacuation of crowded public transport terminals. *Saf. Sci.* **2008**, *46*, 844–857. [CrossRef]

28. Liu, S.; Zhu, G. The application of GIS and IOT technology on building fire evacuation. *Procedia Eng.* **2014**, *71*, 577–582. [CrossRef]

29. Wang, J.; Zhao, H.; Winter, S. Integrating sensing, routing and timing for indoor evacuation. *Fire Saf. J.* **2015**, *78*, 111–121. [CrossRef]

30. Ding, Y.; He, X.; Zhu, Q.; Lin, H.; Hu, M. A dynamic optimization method of indoor fire evacuation route based on realtime situation awareness. *Acta Geod. Cartogr. Sin.* **2016**, *45*, 1464–1475. [CrossRef]

31. Zhu, J.; She, P.; Li, W.; Cao, Y.; Qi, H.; Wang, B.; Wang, Y. Dynamic planning method of indoor fire escape path based on navigation grid. *J. Southwest Jiaotong Univ.* in press. Available online: http://kns.cnki.net/kcms/detail/51.1277.U.20190320.1146.018.html (accessed on 14 January 2020).

32. Han, L.; Guo, H.; Zhang, H. An algorithm for route planning applied in multi-exit indoor emergency. *Sci. Surv. Mapp.* **2018**, *43*, 105–110. [CrossRef]

International Journal of
Geo-Information

Article

A Hybrid Framework for High-Performance Modeling of Three-Dimensional Pipe Networks

Shaohua Wang [1,2], Yeran Sun [3,*], Yinle Sun [4], Yong Guan [5], Zhenhua Feng [4], Hao Lu [4], Wenwen Cai [4] and Liang Long [1]

[1]   Institute of Geographic Sciences and Natural Resources Research, Chinese Academy of Science, Beijing 100101, China; wangshaohua@lreis.ac.cn (S.W.); longl@lreis.ac.cn (L.L.)
[2]   Department of Geography, University of California, Santa Barbara, CA 93117, USA
[3]   School of Geography and Planning, Sun Yat-sen University, Guangzhou 510275, China
[4]   SuperMap Software Co., Ltd., Beijing 100015, China; sunyinle@supermap.com (Y.S.); fengzhenhua@supermap.com (Z.F.); luhao@supermap.com (H.L.); caiwenwen@supermap.com (W.C.)
[5]   Department of Information System and Technology, Claremont Graduate University, Claremont, CA 91711, USA; yong.guan@cgu.edu
*   Correspondence: sunyr8@mail.sysu.edu.cn; Tel.: +86-020-8411-5833

Received: 22 July 2019; Accepted: 29 September 2019; Published: 8 October 2019

Abstract: Three-dimensional (3D) pipe network modeling plays an essential part in high performance-based smart city applications. Given that massive 3D pipe networks tend to be difficult to manage and to visualize, we propose in this study a hybrid framework for high-performance modeling of a 3D pipe network, including pipe network data model and high-performance modeling. The pipe network data model is devoted to three-dimensional pipe network construction based on network topology and building information models (BIMs). According to the topological relationships of the pipe point pipelines, the pipe network is decomposed into multiple pipe segment units. The high-performance modeling of 3D pipe network contains a spatial 3D model, the instantiation, adaptive rendering, and combination parallel computing. Spatial 3D model (S3M) is proposed for spatial data transmission, exchange, and visualization of massive and multi-source 3D spatial data. The combination parallel computing framework with GPU and OpenMP was developed to reduce the processing time for pipe networks. The results of the experiments showed that the hybrid framework achieves a high efficiency and the hardware resource occupation is reduced.

Keywords: three-dimensional (3D) pipe network; building information models (BIM); spatial 3D model (S3M); parallel computing; SuperMap GIS

1. Introduction

With the continuous development and advancement of geographic information systems (GIS), it has become challenging for traditional two-dimensional GIS to meet the needs of geospatial visualizations and analysis [1,2]. With the widespread application of GIS in society, the application of three-dimensional (3D) scenes in smart cities is becoming deeper and wider [3–5]. The traditional two-dimensional management model has been unable to meet the actual needs of the pipe network and pipe big data information analysis, expression, and application. The three-dimensional pipe network occupies a large part of the application scenarios. The performance-based city planning includes data management, data modeling, data analysis and data visualization. Due to the large-scale data size, wide range and various pipeline networks and pipelines, both on the ground and underground, and the spatial distribution is complex and varied, high-performance modeling of 3D pipe networks remains challenging. It is therefore essential to study the 3D pipe network construction and high-performance rendering.

Geodesign is proposed in order to visualize models for city planning [6,7]. With the development of Building Information Modeling (BIM), a three-dimensional model of construction project or facility's physical and functional characteristics, and a shared knowledge has gradually become a common data expression in the field of architecture. Resources provide a decision-making basis for the management of the life cycle of a project or a facility. At different stages of the life cycle, various stakeholders modify information through BIM to coordinate operations. BIM data is an important data source for 3D GIS [8,9]; it enables 3D GIS to move from macro to micro, and at the same time to achieve refined management—especially for pipe networks.

Utility networks are modeled using graph-theory abstractions [10,11], because the structure properties of graph provide network nodes and links. Most utility networks are based on two dimensional modeling design [12], but in some cases, applications such utility networks operate in 3D [13–15]. Pipe network is one type of utility network, the construction of a 3D pipe network is based on network science and GIS. As the BIM-oriented network model can be used in indoor network applications [16], a 3D pipe network should be used in both outdoor and indoor GIS. It is challenging to design 3D pipe networks by considering visualization, checking topology, data management, and 3D network analysis.

The 3D pipe network includes the pipe point portion and the pipeline portion. In the existing three-dimensional systems, the pipeline portion is normally constructed by vector line lofting, and the pipe point is generally built by manual modeling and then imported into the scene. The workflow of these pipe point models requires a lot of manpower and material resources. At the same time, for 3D pipe network data, the user's data source is often vector point and line data. If the model's display mode is adopted, data redundancy and extra workload are unfortunately inevitable. The maintenance of vector seed of point, line data and pipe point models is required as well. In addition, in some enterprise-level scenarios, pipe network data in 3D city models is extremely dense, and thus rendering these dense pipe networks requires a lot of hardware resources. Therefore, improving the carrying capacity and rendering performance of massive pipe network data can be the answer to current capacity and rendering performance limits.

GIS-based methods and modeling are widely used in urban planning [17–21]. GIS and BIM are located in two different sectors for building a smart city. BIM provides the data foundation, and GIS provides a spatial reference, spatial analysis and decision making [22–24]. The combination of BIM and GIS contributes to urban planning, water conservancy projects, railway information modeling, underground pipe network information modeling, ancient building modeling and other fields [25,26]. The integration of GIS and Computer-aided design (CAD) supplies new data sources for building a smart city [27,28]. A hybrid system for expanding 2D GIS into the 3D scene is an effective way to consider the integration of 2D GIS and 3D GIS [29]. There are fewer studies that focus on the integration of spatial data modeling for GIS, BIM, CAD, and oblique photography. It is critical to study the integration model for spatial 3D data.

The computing-intensive framework plays an essential part in high-performance based smart city applications [30]. Parallel computing is an effective way to visualize 3D city models [31,32]. 3D city modeling requires effective algorithms [33,34]. NVIDIA, inventor of the graphics processing unit (GPU), proposes a parallel computing framework based on computing unified device architecture (CUDA), which can be used to speed up the processing efficiency of geospatial applications [35–37]. The multi-threaded parallel processing based on multi-core central processing unit(CPU)s provides extensive computational capabilities [38,39]. The scientific computing for building city modeling is shifting from a CPU-centric central processing model to a collaborative processing model in which CPUs and GPUs work together [40].

In this paper, we propose a hybrid framework for high-performance modeling of 3D pipe networks. This paper is organized as follows: Section 2 introduces a pipe network data model, Section 3 shows hybrid high-performance modeling of 3D pipe networks, Section 4 describes the experiments and the results, and the last section is the conclusion and further work.

2. Pipe Network Data Model

The network data model is an essential part of high-performance modeling of 3D pipe networks.

2.1. Pipe Network Data Structure

A variety of pipe networks are normally represented hierarchically [41]. A network data set is used to describe a type of pipe network information. The data volume is less redundant and contains topology information, which can be used for subsequent spatial network analysis, such as burst analysis and connectivity analysis. Each pipeline data set consists of two sub-datasets: lines and nodes. The line sub-dataset represents pipeline objects, and the node sub-dataset represents point objects, such as elbows, tees, crosses, and valves.

Table 1 shows the main structure of the attribute table of the line sub-dataset, where the ID field is the ID of the line object itself, the FromNode and ToNode fields record the ID of the node object connected to both ends of a line object, and the PrevPoint and NextPoint fields recording the geometric coordinates of the other endpoints of the other line objects connected at both ends can facilitate the subsequent construction of the visualized point model. (x_i, y_i, z_i) is the i-th point coordinate of geometry information. In addition, the geometric coordinates of the line object itself and some attribute information are stored.

Table 1. The pipe line structure in the network data set.

ID Number	From Node	To Node	PrePoint	NextPoint	Geometry Information
1	3	2	(Y_{p_1}, Y_{p_2})	$(Y_{p_{n1}}, Y_{p_{n2}})$	$(x_1, y_1, z_1)(x_2, y_2, z_2)\ldots(x_n, y_n, z_n)$

Table 2 shows the main storage structure of the node sub-dataset, wherein the ID field is the ID of the pipe point object, the ID also corresponds to the FromNode and ToNode fields in the line sub-dataset, and a field for storing geometric information and attribute information. In these two associated data sets, the topology information is stored in the online data set.

Table 2. The node structure in the network data set.

ID Number	Geometry Information
1	(x_1, y_1, z_1)
2	(x_2, y_2, z_2)
3	(x_3, y_3, z_3)
...	...
n	(x_n, y_n, z_n)

2.2. Construction of Pipeline Model

2.2.1. Coordinate Computation of Pipeline Network Model

After building the acquired pipeline data as a network dataset, it will be necessary to create a 3D pipeline network model for rendering. We use the pipe segment corresponding to the line segment between the two coordinates on the line geometry object of the pipeline as a rendering unit, and then combine these pipe segments to form the entire pipeline.

Each pipe segment consists of two sections, and if the position of the vertices on the two sections on the pipe section can be determined, an entire pipeline can be determined. We take the center of each section as the origin, the X-axis along the pipeline, and the Cartesian coordinate system on the Y-axis in the section (as shown in Figure 1). Since the cross-sectional shape of the pipeline is known, the position of the point on the pipe cross section in the local coordinate system can be determined. The position P_s of the center of the section in the world coordinate system is stored in the geometric information of the online object so that the translation matrix (M_t) of the center of the section can be obtained.

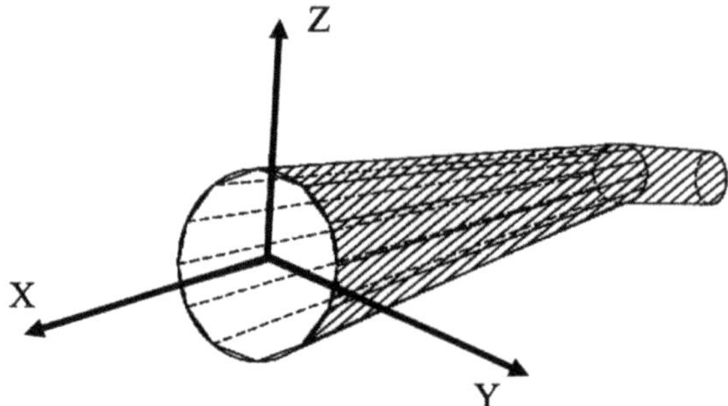

Figure 1. Local coordinate system of pipe section.

To make the section perpendicular to the pipeline, and to make the upper direction of the pipeline section coincide with the upper direction in the world space, a rotation is also required. It is known that the world coordinate pointing to the center point of a given section is W_1, and the coordinate pointing to the center point of the next section along the strike direction is W_2, and the direction of the strike is $\vec{V}_{sec} = W_1 - W_2$, corresponding to the X-axis in the local coordinate system of the section, and the upper direction in the world space is $\vec{V}_{up} = (0, 0, 1)$, corresponding to the Z axis in the local coordinate system and the Y axis being $\vec{V}_r = \vec{V}_{sec} \times \vec{V}_{up}$, the rotation matrix M_{rot} can be obtained. The coordinates of the final section vertices in the world coordinate system are defined as formulation (1).

$$P_w = M_{rot}{\cdot}M_t{\cdot}P_s. \tag{1}$$

It is worth noting that for a longitudinal tube, the up direction is the same or opposite to the direction. In this case, V_r calculated from the previous section can be used as the V_r of the current section, or any direction perpendicular to the direction can be specified as V_r.

2.2.2. Smoothing of Pipeline Inflection Points

When the pipeline is an inflection point, it is necessary to process the cross-section rotation matrix of the connected pipe segments before and after the inflection point so that the front and rear sections can be connected. The inflection point smoothing effect can be achieved by inserting several sections with smooth transitions, while the original two sections need to be retracted a distance to meet the smooth section, as shown in Figure 2.

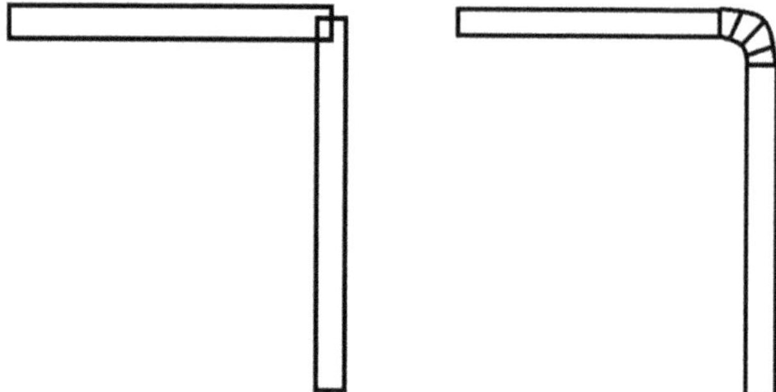

Figure 2. Construction of smooth tube.

In this study, the following algorithm was used to calculate the cross-sectional matrix smoothed by the inflection point: Project the two lines that meet the inflection point into the plane where the two lines are located. Make a circle in the plane tangent to the two lines. As shown in Figure 3, the arc AB is the position of the center-line of the transition section. The more transition sections you take on this section, the smoother the inflection point. The length of the point A to the inflection point O is the length M of the pipeline section that needs to be retracted. By the constraint that the radius of the circle cannot be greater than the width W of the pipeline, we can get the value of M using formulation (2):

$$M = \tan\theta * W. \tag{2}$$

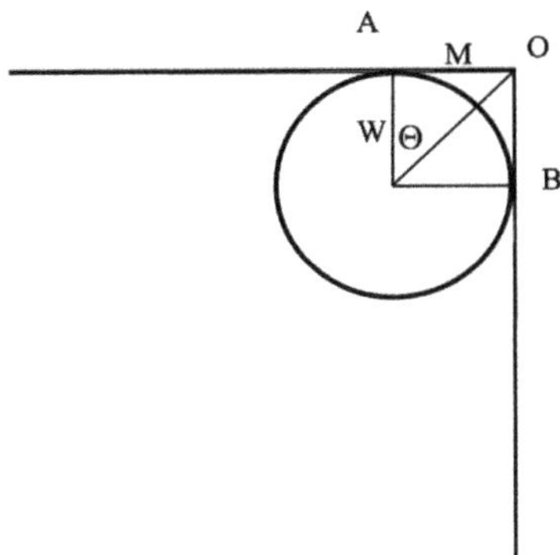

Figure 3. The result of the inflection point projected onto the plane.

The vector OA is defined as $\vec{L}_1$ and OB is defined as vector $\vec{L}_2$. Then, the center coordinate (x_c, y_c, z_c) can be obtained by the following method:

$$\vec{P}_l = \vec{L}_1 \times \vec{L}_2, \tag{3}$$

$$\begin{aligned} k_1 &= \hat{P}_l \cdot \hat{O} \\ k_2 &= \hat{L}_1 \cdot \hat{A} \\ k_3 &= \hat{L}_2 \cdot \hat{B} \end{aligned} \tag{4}$$

$$\begin{aligned} D = &\hat{P}_l.x * \hat{L}_1.y * \hat{L}_2.z + \hat{P}_l.y * \hat{L}_1.z * \hat{L}_2.x + \hat{P}_l.z * \hat{L}_1.x * \hat{L}_2.y \\ &- \hat{P}_l.z * \hat{L}_1.y * \hat{L}_2.x - \hat{P}_l.y * \hat{L}_1.x * \hat{L}_2.z - \hat{P}_l.x * \hat{L}_1.z * \hat{L}_2.y \end{aligned} \tag{5}$$

$$\begin{aligned} x_c = (&k_1 * \hat{L}_1.y * \hat{L}_2.z + k_3 * \hat{P}_l.y * \hat{L}_1.z + k_2 * \hat{P}_l.z * \hat{L}_2.y \\ &- k_3 * \hat{P}_l.z * \hat{L}_1.y - k_2 * \hat{P}_l.y * \hat{L}_2.z - k_1 * \hat{L}_1.z * \hat{L}_2.y)/D \end{aligned} \tag{6}$$

$$\begin{aligned} y_c = (&k_2 * \hat{P}_l.x * \hat{L}_2.z + k_1 * \hat{L}_1.z * \hat{L}_2.x + k_3 * \hat{P}_l.z * \hat{L}_1.x \\ &- k_2 * \hat{P}_l.z * \hat{L}_2.x - k_1 * \hat{L}_1.x * \hat{L}_2.z - k_3 * \hat{P}_l.x * \hat{L}_1.z)/D \end{aligned} \tag{7}$$

$$\begin{aligned} z_c = (&k_3 * \hat{P}_l.x * \hat{L}_1.y + k_2 * \hat{P}_l.y * \hat{L}_1.x + k_1 * \hat{L}_1.x * \hat{L}_2.y \\ &- k_1 * \hat{L}_1.y * \hat{L}_2.x - k_3 * \hat{P}_l.y * \hat{L}_1.x - k_2 * \hat{P}_l.x * \hat{L}_2.y)/D \end{aligned} \tag{8}$$

From the coordinates of the center of the circle, the position of the transition section inserted in the arc can be easily obtained, and then the translation and rotation matrix of the section of the transition section can be calculated by the previous method.

2.3. Construction of the Pipe Point Model

Pipe points such as elbows, tees, and crosses are essential parts of the pipeline scenario. In the previous pipeline information management systems, such a pipe point is normally constructed by a method manually and created in the modeling software, and then the modeled pipe point is imported into the scene. There are two main issues in the way of constructing the pipe point. The first is that the construction of the pipe point model is carried out in its own local coordinate system. Therefore, after the pipe point is imported into the scene, it is necessary to manually adjust its position direction so that the pipe points can be matched with the pipeline. The second is that due to the accuracy of the pipeline data collection, the angle between the pipe and the pipe point is different from the standard value. Therefore, the standard three-way and four-way models cannot be used to match the pipeline. It is necessary to construct a pipe point model for each angle, which greatly increases the workload of modeling.

2.3.1. Constructing the Pipe Point Model

We propose a new method for constructing the pipe point model, which can quickly construct the elbows, tees, and crosses that match the connected pipelines. Among them, the elbow is a pipe point connected to two pipes. The following paragraph is an example of the construction of such a pipe point connected to two or more pipes by taking a multi-pass pipe point as an example.

A multi-pass point can be broken down into a combination of multiple half-pipe segments, as shown in Figure 4. The shape of each half pipe segment is determined by the center point of the pipe point and the direction of the pipe connected to it, so that one half-pipe segment becomes HalfPipePair. Its structure contains three points including first point (FirstPt), center point (CenterPt), and second point (SecondPt).

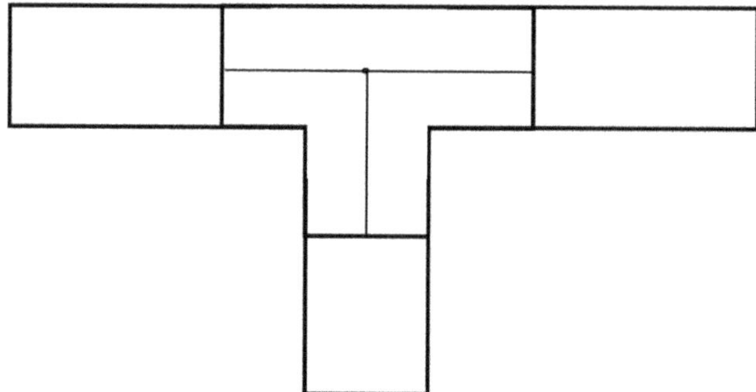

Figure 4. The division of multi-way pipe.

We define that the half pipe pair is made up of a section on the X > 0 side of the pipeline, and its direction is FirstPt->CenterPt->SecondPt.

Use the following algorithm to find the half pipe pairs needed to make up the multi-pass point.

(1) Traverse the points A, B on each of the two pipelines, calculate their plane P with the center point O;

(2) Traverse the other pipeline points C except for A and B, calculate the position it projects onto P, and record whether the half pipe pairs of $A \rightarrow O \rightarrow B$ and the half pipe pairs of $B \rightarrow O \rightarrow A$ are required as the boolean variables *bneedAtoB* and *bneedBtoA*. The initial value is true;

(3) Determine the relationship between C_p and $\angle AOB$. If C_p falls inside $\angle AOB$, then *bneedAtoB = false*; if C_p falls outside $\angle AOB$, then *bneedBtoA = false*;

(4) After traversing the other points except for A and B, if the *needAtoB* is true, the half pipe pair $A \rightarrow O \rightarrow B$ is recorded. If the *bneedBtoA* is *true*, the half pipe pair $B \rightarrow O \rightarrow A$ is recorded; back to (1).

2.3.2. Matrix Computation of Half-Pipe

After a set of half-pipe pairs constituting the multi-pass pipe point are obtained, they need to be decomposed into pipe segment units. As shown in Figure 5, one half pipe pair can be broken down into two pipe segment units. The calculation of the section matrix that is connected to the pipeline is given in Section 2.2.1. The matrix calculation of the intersection of two pipe segments is described below. It is easy to know that the interface is the angle bisector of $\angle AOB$, and the trend is $\vec{V}_{sec} = \overrightarrow{AO} + \overrightarrow{OB}$. Therefore, its rotation matrix can be calculated by the method in Section 2.2.1.

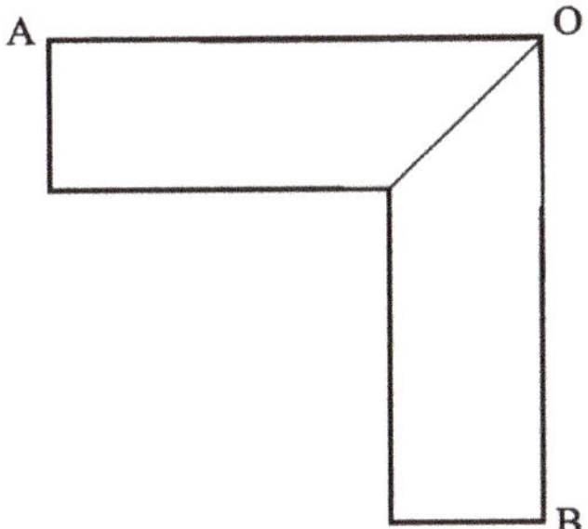

Figure 5. The plane projection of half tube.

Let $\angle AOB$ be α, the normal vector of AOB plane be $\vec{P}$, and the angle between $\vec{P}$ and upper direction $\vec{V}_{up}$ be β. Then, the scaling ratios $Scale_Y$ and $Scale_Z$ of the Y and Z axes in the local coordinate system and the scaling matrix M_{Scale} can be calculated. The vertex position on the half pipe segment is $P_W = M_{Scale} \cdot M_{rot} \cdot M_t \cdot P_s$.

$$Scale_Y = \sqrt{\sin^2 \beta / \sin^2(\alpha * 0.5) + \cos^2 \beta}, \tag{9}$$

$$Scale_Z = \sqrt{\cos^2 \beta / \sin^2(\alpha * 0.5) + \sin^2 \beta}, \tag{10}$$

$$M_{Scale} = \begin{pmatrix} 1 & 0 & 0 & 0 \\ 0 & Scale_Y & 0 & 0 \\ 0 & 0 & Scale_Z & 0 \\ 0 & 0 & 0 & 1 \end{pmatrix}. \tag{11}$$

2.4. Integration Framework for 2D GIS and 3D GIS

The integration framework for 2D GIS and 3D GIS contains spatial data management integration, application integration, function module integration, expression symbol system integration, and analysis function integration. The essential part is that the spatial data model and spatial data structure of 2D and 3D data objects are kept in the same design. All the two-dimensional data can be directly used with high-performance modeling in the 3D scene without any spatial data conversion. 2D maps and 3D scenes can easily be generated based on the integration framework for 2D GIS and 3D GIS (Figure 6). The 2D and 3D integration of spatial data management solves the problem of spatial data compatibility of different dimensions and reduces the cost and complexity of the system construction to meet the needs of various applications. Spatial analysis is supported in both 2D map and 3D scene. 3D network analysis is widely used in pipeline network management. GPU graphics hardware acceleration provides a powerful support for 3D analysis functions and brings users a good experience with high performance.

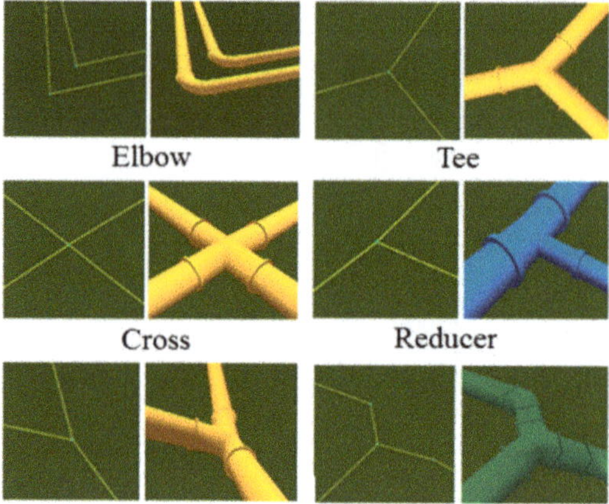

Figure 6. Pipe network models in 2D GIS and 3D GIS.

3. High-Performance Modeling for 3D Pipe Networks

The framework for high-performance modeling of 3D pipe network contains a spatial 3D model, instantiation, adaptive rendering, and combinational parallel computing.

3.1. Spatial 3D Model (S3M)

Spatial 3D model (S3M) is proposed for spatial data transmission, exchange and high-performance visualization of massive and multi-source 3D spatial data [42]. It meets the requirements of users in three-dimensional data transmission and analysis between different terminals (mobiles, browsers, and desktops) through an efficient and scalable data specification. S3M supports multi-source data, including pipelines, BIM, laser point clouds, vector, terrain, dynamic water surfaces, and 3D grids. S3M provides the ability to efficiently visualize a large amount of data using the level of details (LOD), batch rendering, and instantiation to improve the rendering performance.

The S3M includes description files, index tree files, data files, and attribute files. The file organization of each object storage is shown in Table 3. S3M TileTreeSet object is the basic element in S3M. TileTreeSetInfo is its description information, which is an overall description of the data. If the TileTreeSet is constructed based on the point, line, polygon or model dataset, there may be an AttributeInfos that represents attribute description information for each data set. Space division of three-dimensional data are in a specified spatial range, each spatial division corresponding to a tree structure organization tile collection is represented by a TileTree object. IndexTree is index information of its tree structure. AttributeData representing attribute data are recorded in TileTree attribute data for each object. Each TileTree is subdivided step by step from top to bottom, and each space partition corresponds to one tile, represented by a Tile object.

Table 3. The file organization of S3M.

Object	Storage Format	File Type	Description
TileTreeSetInfo	.scp	description file	description of the entire data
AttributeInfos	attribute .xml	attribute description file	description of each data set attribute in TileTreeSet
TileTree	Folder	data folder	store all data in the tile range
AttributeData	.xml	attribute data file	attribute data of all objects belong to the tile
IndexTree	.json	index tree file	all PagedLOD information belong to the tile
Tile	.s3mb	data file	a S3MB file stores data in a spatial division of the LOD layer

The description file (.scp) and data folder are the basic components. The description file contains the path filename (.json) for each TileTree. The index file is a description of the tree structure of the tile data and can obtain the bounding box of the tile file of each layer, the switching information of the LOD and the attached child node file without loading the actual data. The main role is to accelerate the efficiency of tile file retrieval; the attribute data includes one attribute description file (attribute.xml) and a .xml file storing each tile attribute data in each TileTree. All tiles in TileTree form a tree-like logical structure.

To facilitate sharing of multi-source heterogeneous geospatial data, the open geospatial database connectivity (OGDC) was proposed to connect different types of geospatial data (Figure 7). OGDC provides a unified entry to geospatial data [43]. The unified data specifications, such as I3S [18] and S3M, can be used for high-performance, consistent access to data in the system, such as various GIS terminal applications, and user APPs.

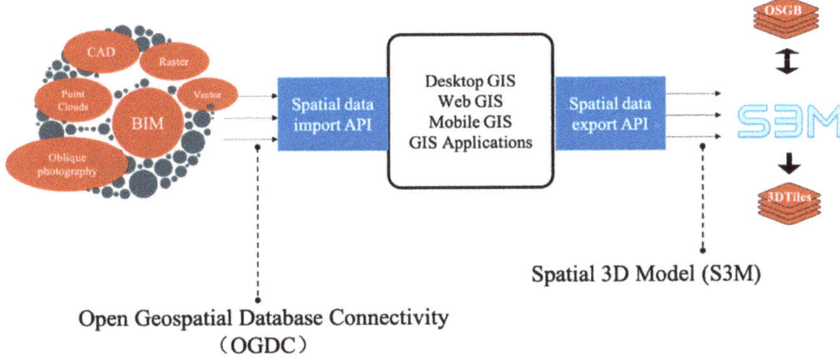

Figure 7. The heterogeneous geospatial data fusion framework.

3.2. Instantiation Rendering of Pipe Networks

Instantiated rendering refers to multiple renderings of a grid using different parameters at different locations. Instantiation is often applied to the rendering of static mesh objects such as leaves, grass, etc. in a large number of scenes. All instances share a vertex buffer, which is used to store a single grid of data that will be instantiated multiple times. The instance buffer is used to store instance data for each object, including the information such as transformation matrices, color data, and lighting data. The rendering process is combined with a vertex buffer and an instance buffer for rendering.

The pipeline is in fact composed of pipe sections that are substantially similar and have slightly different positions and directions. In order to render these slightly different pipe segments, the traditional method is to store each pipe segment separately, which consumes a lot of memory resources; or it needs to switch frequently between rendering states, such as rotating and panning each pipe segment, which significantly affects rendering performance. By instantiating the grid of reused pipe segments, reducing the number of calls and memory requirements, and giving most of the rendering work to the GPU, reducing the CPU load is a good way to improve rendering performance.

For the instantiation of the pipe point pipeline, only two sets of grids need to be established. One set is a complete section pipe section for rendering the elbows in the pipeline and the pipe points, and the other set is a half-section pipe section for rendering the multi-pass pipe points. As shown in Figure 8, the different tube segment examples differ in the position of the cross-section at both ends, in the size of the zoom, and in the direction. These can be represented by two transformation matrices M_{Bottom} and M_{Top}. The matrix stores 18 floating point numbers representing the three rows and three columns of the two matrices into the instance buffer, and the grid data of the pipe segment is stored in the vertex buffer. In addition, the vertex needs to be stored in the vertex attribute of the vertex buffer, to which the matrix belongs before and after. Finally, the vertex position of the pipeline can be calculated by multiplying the vertices of the static mesh with the matrix in the vertex shader.

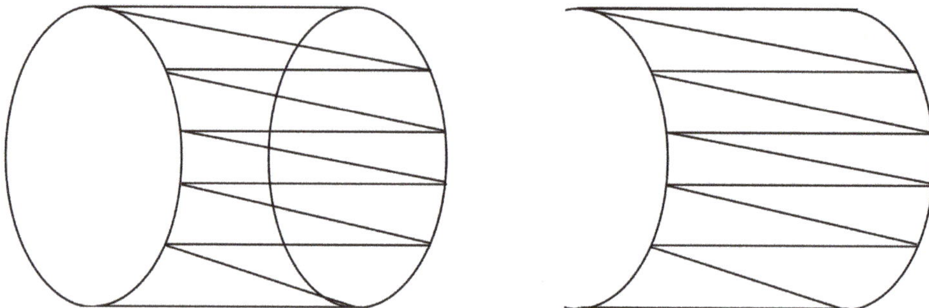

Figure 8. The grid of section and half cross section.

3.3. Adaptive Rendering for Pipe Network Data

The BIM data structure includes a spatial data model and attributes data. The spatial data model includes spatial information and appearance shape. The attribute data includes design parameters, construction parameters and operation, and maintenance parameters. The 3D GIS covers the data structure of BIM (spatial data and attribute data), covering the data representation of BIM (3D model), covering BIM data objects (BIM for architectural objects, and GIS covers a wide range, including architectural objects), and it overlaps with the BIM function (information management and spatial analysis). The 3D underground pipeline scene based on BIM-GIS consists of three-dimensional pipe points and three-dimensional pipelines. The three-dimensional pipeline includes round pipes, square grooves, pipe blocks, and vertical pipes. Three-dimensional pipe points include three types of feature points, wells and ancillary facilities: feature points such as elbows, straight throughs, three-way, cover plugs, and pipe caps. Wells include square wells, round wells, and rain rafts. Ancillary facilities include valves and water meters. Based on the BIM technology, this study uses linear symbols to construct three-dimensional pipelines and adaptive pipe point symbols to construct three-dimensional pipe points. Some special feature points, wells, and ancillary facilities are displayed by model symbols.

BIM model symbols can be used to visualize the pipe points with complex shapes, such as valves. However, in the three-dimensional scene, the model symbols cannot adaptively adjust the angle according to the direction of the pipeline in the X, Y, and Z directions. Due to the influence of the model angle, the pipe point deviates from the scene, for example, the pipe at the bottom of the valve cannot be connected with the pipeline and the valve switch is covered by a pipe. We can convert the BIM model to 3D pipe network using S3M (Figure 9).

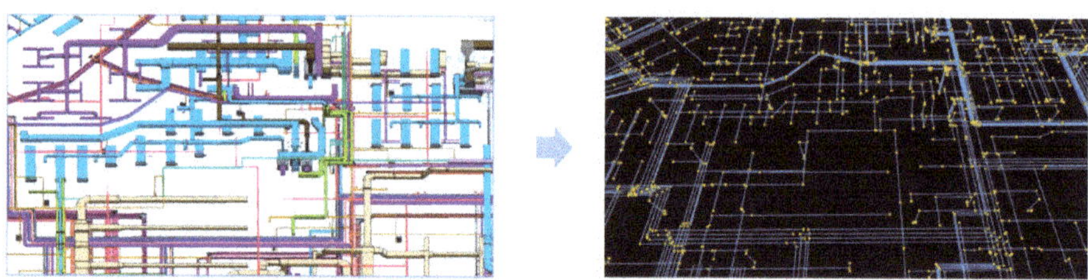

Figure 9. From BIM model to 3D pipe network data model.

An adaptive rendering model is proposed with the combinations of the data structure of the BIM model and the spatial structure characteristics of 3D GIS. The adaptive rendering model contains a pipe layer setting method and a symbol matching method. The pipe point layer was shown in a custom thematic map, specifying a property field as the symbol style of the thematic map. After setting the parameters of the symbol model, the symbol matching method adjusts the angle according to the pipeline orientation, including the pipe at the bottom of the valve in the pipeline direction, and the valve switch perpendicular to the pipeline direction (Figure 10).

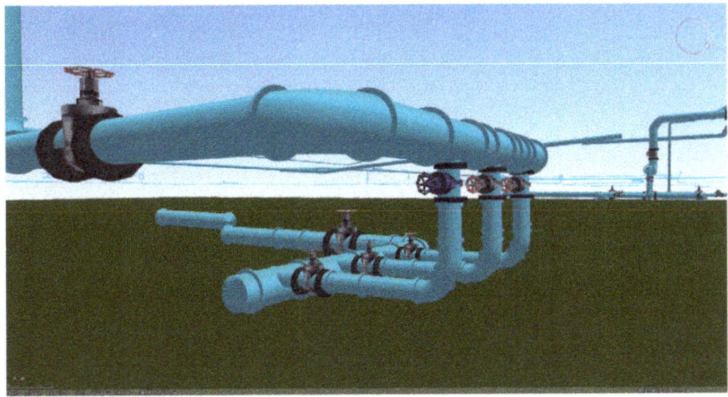

Figure 10. The adaptive rendering map in 3D GIS.

3.4. Combination Computing Framework with GPU and OpenMP

Rendering large-scale 3D pipe networks requires intensive computational time. The combinational parallel computing framework with GPU and OpenMP can greatly improve computing intensive pipe network-based applications. Considering that the spatial data exchange between the memory and the memory is time-consuming, and the large pipe network data cannot be copied to video memory at one time. We used the method of reading the pipe network data into the video memory. At the same time, we introduce the multi-thread processing strategy based on OpenMP to minimize the time occupation of GPU computing and data exchange to achieve the purpose of hiding data exchange time as much as possible.

Each CUDA core is only responsible for calculating the pixel result value of one specific location at the same time, and the kernel function specifies the specific calculation method. The specific workflow of a kernel function is to obtain the location of the pixel according to the built-in thread ID, and obtain the neighboring pixel at the location of the pixel (Figure 11). Then, the cell value is calculated based on the specific operator. Among them, the temporary variable corresponding to the neighboring pixel value is stored by using the shared memory inside the CUDA thread block, which can effectively improve the analysis performance. The computational resources were recycled when the combinational parallel computing framework was finished.

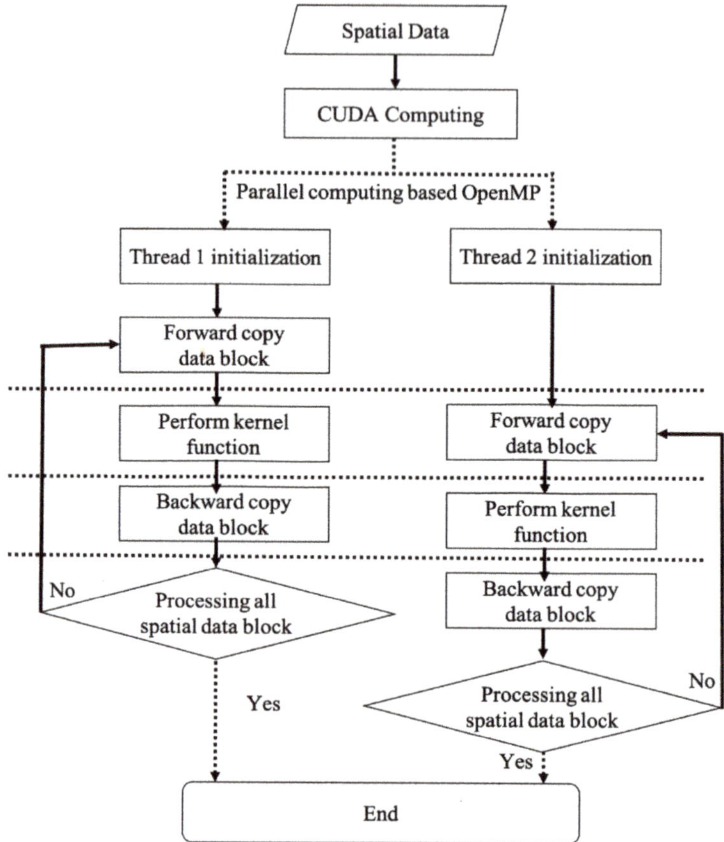

Figure 11. The workflow of the combinational parallel computing framework with GPU and OpenMP.

4. Experiments

The 3D pipe network model was implemented and the performance of the visualization is verified. The open source GIS projects are developed to verify the efficiency for the hybrid method of modeling the 3D pipe network. 3D pipe network model can be shown in desktop GIS, mobile GIS, and web GIS. The iDesktop-cross [44] is an open source desktop GIS project [45], which includes 3D pipe network models. The SuperMap iClient GIS (iClient-JavaScript [46]) is a visual analytics framework for WebGIS based application [7]. The pipe network data is shown in the outdoor GIS (Figure 12a) and the indoor GIS (Figure 12b).

a) Pipe network model in the outdoor GIS.

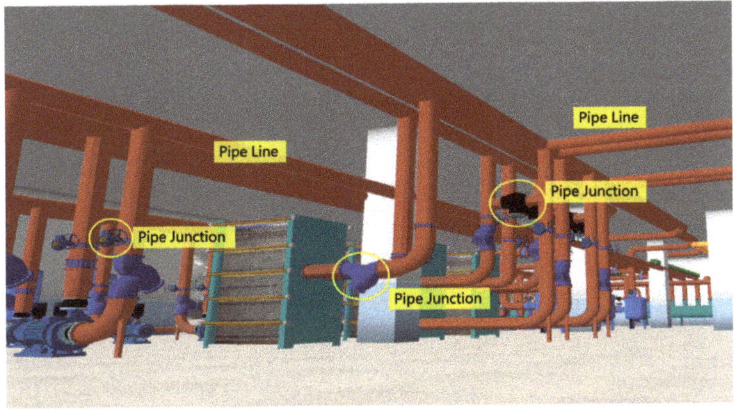

b) Pipe network model in the indoor GIS.

Figure 12. Pipeline network model in iDesktop Cross.

Based on the API provided by iClient3D for WebGL [47], the pipe network visualization is shown in Figure 13 and is developed by SuperMap iEarth for WebGL [48].

To test the performance using the methods above, we use the instantiation and non-instantiation techniques to test the performance of the applications of 3D pipeline networks. The experimental environment is in the 64-bit Windows 7 operating system with 4 GB memory, the graphics card uses GTX650 with 2 G memory, and the CPU is an Intel i5-3340 with 3.1 GHz quad-core processor. The test scenario is pipeline data of a certain plant area, the area is about 5 km^2, and the pipeline is densely distributed, as shown in Figure 14. We specify the flight route in the scene, then let the camera automatically move along the flight path, recording the frame rate, CPU usage, and memory usage at each moment along the way.

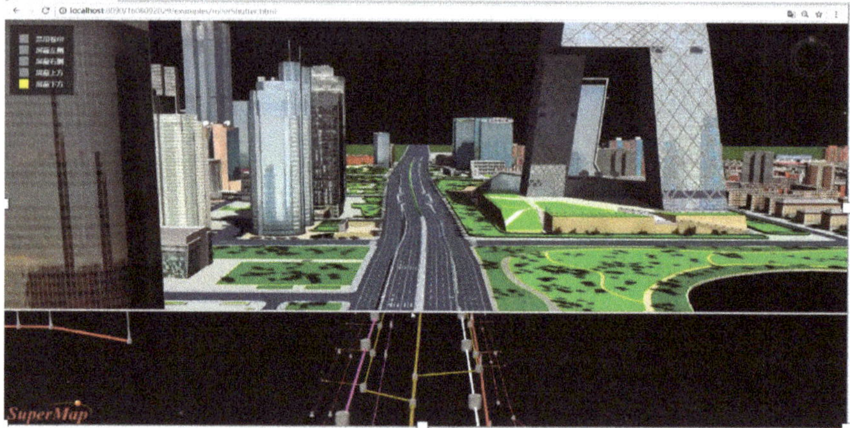

Figure 13. Pipeline network in the SuperMap iEarth.

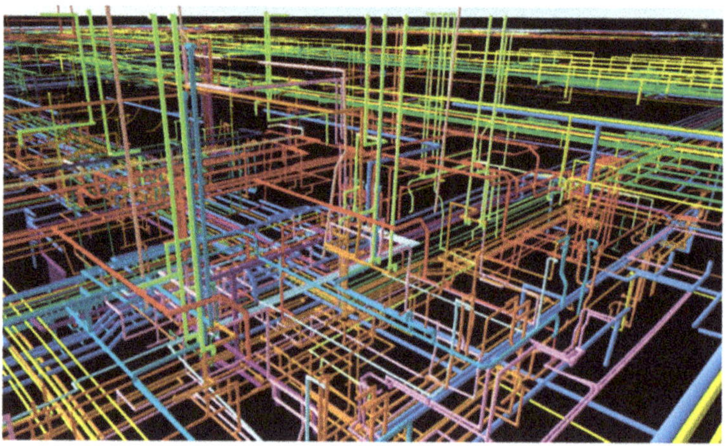

Figure 14. Pipeline network in the experiment.

The experimental results are shown in Table 4. By using the instantiation method, the frame rate is increased by about 100%, the CPU usage is reduced to 33%, the memory usage is reduced to 25%, and the memory usage is slightly reduced. Proving that instantiation technology improves pipeline rendering performance is significant.

Table 4. The result of benchmark experiment.

Experiment Type	Rendering Frame Rate	Percentage of CPU (%)	Physical Memory (MB)	Video Memory (MB)
Non-instantiation method	32.44	32.29	580.39	340-550
Instantiation method	61.66	9.34	552.57	310-360

5. Conclusions and Discussion

In summary, we propose a hybrid framework for high-performance modeling of 3D pipe network, including pipe network data model and high-performance modeling. The pipe network is decomposed into two sections described in Section 2, and the multi-pass pipe point is decomposed into a set of pipe segments described by two half sections, through the topological relationship between the pipe point and the pipeline. Through this decomposition, complex pipe points can be split into simple units. In this way, a well-matched pipe point pipeline model can be quickly constructed, saving a lot of manual operations and improving the efficiency of building pipeline scenarios. In addition, the split pipe segment unit can be easily combined with the instantiation technique for rendering.

The results of the experiments have shown that the use of instantiation technology significantly improves the rendering performance of the 3D pipe networks. 3D pipe network design needs more time. The hybrid method reduces the cost for constructing 3D pipe network scene and improves the rendering performance of 3D pipelines and 3D pipe points. The geospatial application of 3D pipe network is complex. Spatial 3D model (S3M) is proposed for spatial data transmission, exchange, and visualization of massive and multi-source 3D spatial data. Rendering large-scale 3D pipe networks requires intensive computational resources. The combination parallel computing framework with GPU and OpenMP significantly reduces the processing time for large-scale 3D pipe networks. The results of the experiments showed that the hybrid framework achieves a high efficiency and the hardware resource occupation is reduced.

The hybrid framework for high-performance modeling of 3D pipe network that integrates geospatial applications makes GIS integrate into the high-performance based smart city with unprecedented opportunities. There are however some limitations to this study. (1) This study did not consider spatial cloud computing and edge computing, which is more powerful than hybrid parallel computing. Spatial cloud computing can address the high-performance based challenges for large-scale geospatial applications because spatial cloud computing has the ability to process large-scale 3D GIS models with high performance. Edge computing enables spatial analytics and geospatial data gathering to process near the source of the spatial data. Edge computing has the opportunity to perform real-time 3D pipe network modeling. (2) Our models did not consider artificial intelligence (AI) properties. 3D pipe network data production takes a long time with manual work. It is essential to extract the pipe network data from LiDAR data and remote sensing images using AI-based methods. In future work, we will focus on the integration of spatial cloud computing, edge computing, and machine learning to high-performance based smart city applications.

Author Contributions: Conceptualization, Shaohua Wang; data curation, Yinle Sun, Wenwen Cai and Liang Long; funding acquisition, Shaohua Wang and Liang Long; investigation, Shaohua Wang and Yeran Sun; methodology, Shaohua Wang and Hao Lu; project administration, Zhenhua Feng; software, Yinle Sun, Zhenhua Feng and Hao Lu; supervision, Shaohua Wang and Zhenhua Feng; validation, Wenwen Cai and Yong Guan; visualization, Yinle Sun and Yong Guan; writing—original draft, Shaohua Wang and Yeran Sun.

Funding: This research was funded by the Fundamental Research Funds for the Central Universities of China (Grant No. 37000-18841208), the National Key R&D Plan (2016YFB0502004), National Postdoctoral International Exchange Program (Grant Number 20150081), Project of Beijing Excellent Talents (201500002685XG242), Independent Research Project of State Key Laboratory of Resources and Environmental Information Systems, Chinese Academy of Sciences (088RAC00YA).

Acknowledgments: We would like to acknowledge the Institute of Geographic Sciences and Natural Resources Research of the Chinese Academy of Science for providing a research grant to conduct this work.

Conflicts of Interest: The authors declare no conflict of interest.

References

1. Zhong, E. Geocontrol and live geography: Some thoughts on the direction of gis. *J. Geo-Inf. Sci.* **2013**, *15*, 783–792.
2. Anejionu, O.; Thakuriahab, P.; McHugh, A.; Sun, Y.; McArthur, D.; Mason, P.; Walpol, R. Spatial urban data system: A cloud-enabled big data infrastructure for social and economic urban analytics. *Future Gener. Comput. Syst.* **2019**, *98*, 18. [CrossRef]
3. Biljecki, F.; Ledoux, H.; Stoter, J. An improved lod specification for 3D building models. *Comput. Environ. Urban Syst.* **2016**, *59*, 25–37. [CrossRef]

4. Wang, S.; Zhong, Y.; Wang, E. An integrated gis platform architecture for spatiotemporal big data. *Future Gener. Comput. Syst.* **2019**, *94*, 160–172. [CrossRef]

5. Yu, M.; Yang, C. A 3D multi-threshold, region-growing algorithm for identifying dust storm features from model simulations. *Int. J. Geogr. Inf. Sci.* **2017**, *31*, 939–961. [CrossRef]

6. Wu, C.-L.; Chiang, Y.-C. A geodesign framework procedure for developing flood resilient city. *Habitat Int.* **2018**, *75*, 78–89. [CrossRef]

7. Wang, S.; Zhong, E.; Cai, W.; Zhou, Q.; Lu, H.; Gu, Y.; Yun, W.; Hu, Z.; Long, L. A visual analytics framework for big spatiotemporal data. In Proceedings of the 2nd ACM SIGSPATIAL Workshop on Analytics for Local Events and News, Seattle, WA, USA, 6 November 2018; ACM: New York, NY, USA; p. 3.

8. Borrmann, A.; Kolbe, T.H.; Donaubauer, A.; Steuer, H.; Jubierre, J.R.; Flurl, M. Multi-scale geometric-semantic modeling of shield tunnels for gis and bim applications. *Comput. Aided Civ. Infrastruct. Eng.* **2015**, *30*, 263–281. [CrossRef]

9. Amirebrahimi, S.; Rajabifard, A.; Mendis, P.; Ngo, T. A framework for a microscale flood damage assessment and visualization for a building using bim–gis integration. *Int. J. Digit. Earth* **2016**, *9*, 363–386. [CrossRef]

10. Bakalov, P.; Hoel, E.G.; Kim, S. A network model for the utility domain. In Proceedings of the 25th ACM SIGSPATIAL International Conference on Advances in Geographic Information Systems, Redondo Beach, CA, USA, 7–10 November 2017; ACM: New York, NY, USA; p. 32.

11. Oliver, D.; Hoel, E.G. A trace framework for analyzing utility networks: A summary of results (industrial paper). In Proceedings of the 26th ACM SIGSPATIAL International Conference on Advances in Geographic Information Systems, Seattle, WA, USA, 6–9 November 2018; ACM: New York, NY, USA; pp. 249–258.

12. Alam, S.N.; Haas, Z.J. Coverage and connectivity in three-dimensional underwater sensor networks. *Wirel. Commun. Mob. Comput.* **2008**, *8*, 995–1009. [CrossRef]

13. Hijazi, I.; Ehlers, M.; Zlatanova, S.; Isikdag, U. Ifc to citygml transformation framework for geo-analysis: A water utility network case. In Proceedings of the 4th International Workshop on 3D Geo-Information, Ghent, Belgium, 4–5 November 2009.

14. Hijazi, I.; Ehlers, M.; Zlatanova, S.; Becker, T.; van Berlo, L. Initial investigations for modeling interior utilities within 3d geo context: Transforming ifc-interior utility to citygml/utilitynetworkade. In *Advances in 3D Geo-Information Sciences*; Springer: Berlin/Heidelberg, Germany, 2011; pp. 95–113.

15. Becker, T.; Nagel, C.; Kolbe, T.H. Integrated 3d modeling of multi-utility networks and their interdependencies for critical infrastructure analysis. In *Advances in 3D Geo-Information Sciences*; Springer: Berlin/Heidelberg, Germany, 2011; pp. 1–20.

16. Teo, T.-A.; Cho, K.-H. Bim-oriented indoor network model for indoor and outdoor combined route planning. *Adv. Eng. Inform.* **2016**, *30*, 268–282. [CrossRef]

17. Li, X.; Yeh, A.G.-O. Modelling sustainable urban development by the integration of constrained cellular automata and gis. *Int. J. Geogr. Inf. Sci.* **2000**, *14*, 131–152. [CrossRef]

18. Su, D.Z. Gis-based urban modelling: Practices, problems, and prospects. *Int. J. Geogr. Inf. Sci.* **1998**, *12*, 651–671. [CrossRef] [PubMed]

19. Moghadam, S.T.; Toniolo, J.; Mutani, G.; Lombardi, P. A gis-statistical approach for assessing built environment energy use at urban scale. *Sustain. Cities Soc.* **2018**, *37*, 70–84. [CrossRef]

20. Zhang, X.; Ma, G.; Jiang, L.; Zhang, X.; Liu, Y.; Wang, Y.; Zhao, C. Analysis of spatial characteristics of digital signage in beijing with multi-source data. *ISPRS Int. J. Geo-Inf.* **2019**, *8*, 207. [CrossRef]

21. Liu, K.; Gao, S.; Lu, F. Identifying spatial interaction patterns of vehicle movements on urban road networks by topic modelling. *Comput. Environ. Urban Syst.* **2019**, *74*, 50–61. [CrossRef]

22. Song, Y.; Wang, X.; Tan, Y.; Wu, P.; Sutrisna, M.; Cheng, J.; Hampson, K. Trends and opportunities of bim-gis integration in the architecture, engineering and construction industry: A review from a spatio-temporal statistical perspective. *ISPRS Int. J. Geo-Inf.* **2017**, *6*, 397. [CrossRef]

23. Zhu, J.; Wright, G.; Wang, J.; Wang, X. A critical review of the integration of geographic information system and building information modelling at the data level. *ISPRS Int. J. Geo-Inf.* **2018**, *7*, 66. [CrossRef]

24. Kang, T.W.; Hong, C.H. A study on software architecture for effective bim/gis-based facility management data integration. *Autom. Constr.* **2015**, *54*, 25–38. [CrossRef]

25. Liu, X.; Wang, X.; Wright, G.; Cheng, J.; Li, X.; Liu, R. A state-of-the-art review on the integration of building information modeling (bim) and geographic information system (gis). *ISPRS Int. J. Geo-Inf.* **2017**, *6*, 53. [CrossRef]

26. Karan, E.P.; Irizarry, J.; Haymaker, J. Bim and gis integration and interoperability based on semantic web technology. *J. Comput. Civ. Eng.* **2015**, *30*, 04015043. [CrossRef]

27. Bansal, V. Integrated cad and gis–based framework to support construction planning: Case study. *J. Archit. Eng.* **2017**, *23*, 05017005. [CrossRef]

28. Quan, S.J.; Li, Q.; Augenbroe, G.; Brown, J.; Yang, P.P.-J. Urban data and building energy modeling: A gis-based urban building energy modeling system using the urban-epc engine. In *Planning Support Systems and Smart Cities*; Springer: Berlin/Heidelberg, Germany, 2015; pp. 447–469.

29. Yu, L.-J.; Sun, D.-F.; Peng, Z.-R.; Zhang, J. A hybrid system of expanding 2D gis into 3D space. *Cartogr. Geogr. Inf. Sci.* **2012**, *39*, 140–153. [CrossRef]

30. Hu, M.; Li, C. Design smart city based on 3S, internet of things, grid computing and cloud computing technology. In *Internet of Things*; Springer: Berlin/Heidelberg, Germany, 2012; pp. 466–472.

31. Biljecki, F.; Stoter, J.; Ledoux, H.; Zlatanova, S.; Çöltekin, A. Applications of 3d city models: State of the art review. *ISPRS Int. J. Geo-Inf.* **2015**, *4*, 2842–2889. [CrossRef]

32. Liang, J.; Gong, J.; Zhou, J.; Ibrahim, A.N.; Li, M. An open-source 3D solar radiation model integrated with a 3D geographic information system. *Environ. Model. Softw.* **2015**, *64*, 94–101. [CrossRef]

33. Luo, F.; Zhong, E.; Cheng, J.; Huang, Y. Vgis-collide: An effective collision detection algorithm for multiple objects in virtual geographic information system. *Int. J. Digit. Earth* **2011**, *4*, 65–77. [CrossRef]

34. Luo, F.; Zhong, E.; Cao, G.; Tellez, R.D.; Gao, P. Vgis-antijitter: An effective framework for solving jitter problems in virtual geographic information systems. *Int. J. Digit. Earth* **2013**, *6*, 28–50. [CrossRef]

35. Ortega, L.; Rueda, A. Parallel drainage network computation on cuda. *Comput. Geosci.* **2010**, *36*, 171–178. [CrossRef]

36. Wu, J.; Deng, L.; Paul, A. 3d terrain real-time rendering method based on cuda-opengl interoperability. *IETE Tech. Rev.* **2015**, *32*, 471–478. [CrossRef]

37. Zhang, J.; You, S. Cudagis: Report on the design and realization of a massive data parallel gis on gpus. In Proceedings of the 3rd ACM SIGSPATIAL International Workshop on GeoStreaming, Redondo Beach, CA, USA, 6 November 2012; ACM: New York, NY, USA; pp. 101–108.

38. Zhou, C.; Chen, Z.; Pian, Y.; Xiao, N.; Li, M. A parallel scheme for large-scale polygon rasterization on cuda-enabled gpus. *Trans. GIS* **2017**, *21*, 608–631. [CrossRef]

39. Zhang, X.; Huang, C.; Min, G.; Wu, Y.; Zuo, Y. Distributed machine learning in big data era for smart city. In *From Internet of Things to Smart Cities*; Chapman and Hall/CRC: Boca Raton, FL, USA, 2017; pp. 151–177.

40. Heitzler, M.; Lam, J.C.; Hackl, J.; Adey, B.T.; Hurni, L. Gpu-accelerated rendering methods to visually analyze large-scale disaster simulation data. *J. Geovis. Spat. Anal.* **2017**, *1*, 3. [CrossRef]

41. Perelman, L.S.; Allen, M.; Preis, A.; Iqbal, M.; Whittle, A.J. Multi-level automated sub-zoning of water distribution systems. In Proceedings of the 7th Intl. Congress on Env. Modelling and Software, San Diego, CA, USA, 15–19 June 2014.

42. SuperMap. Spatial 3d Model Project. Available online: https://github.com/SuperMap/s3m-spec (accessed on 20 July 2019).

43. SuperMap. The Open Geospatial Database Connectivity Project. Available online: https://github.com/SuperMap/OGDC (accessed on 20 July 2019).

44. SuperMap. The Idesktop-Cross Project. Available online: https://supermap-idesktop.github.io/SuperMap-iDesktop-Cross/ (accessed on 20 July 2019).

45. Cai, W.; Wang, S.; Zhong, E.; Hu, C.; Liu, X. Design and implementation of a new cross-platform open source gis desktop software. *Bull. Surv. Mapp.* **2017**, *1*, 122–125.

46. SuperMap. The Iclient-Javascript Project. Available online: https://github.com/SuperMap/iClient-JavaScript (accessed on 20 July 2019).

47. SuperMap. The Iclient3d for Webgl Project. Available online: https://enonline.supermap.com/SuperMap_iClient3D_9.0.1_for_WebGL/ (accessed on 20 July 2019).

48. SuperMap. Supermap Iearth for Webgl Project. Available online: http://www.supermapol.com/earth/ (accessed on 20 July 2019).

International Journal of
Geo-Information

MDPI

Article

Direction-Aware Continuous Moving *K*-Nearest-Neighbor Query in Road Networks

Tianyang Dong [1,*]**, Lulu Yuan** [1]**, Yuehui Shang** [1]**, Yang Ye** [1] **and Ling Zhang** [2]

[1]   College of Computer Science and Technology, Zhejiang University of Technology, Hangzhou 310023, China
[2]   School of Economics and Management, Zhejiang University of Science and Technology, Hangzhou 310023, China
*   Correspondence: dty@zjut.edu.cn

Received: 29 May 2019; Accepted: 21 August 2019; Published: 29 August 2019

Abstract: Continuous *K*-nearest neighbor (CKNN) queries on moving objects retrieve the *K*-nearest neighbors of all points along a query trajectory. They mainly deal with the moving objects that are nearest to the moving user within a specified period of time. The existing methods of CKNN queries often recommend *K* objects to users based on distance, but they do not consider the moving directions of objects in a road network. Although a few CKNN query methods consider the movement directions of moving objects in Euclidean space, no efficient direction determination algorithm has been applied to CKNN queries over data streams in spatial road networks until now. In order to find the top *K*-nearest objects move towards the query object within a period of time, this paper presents a novel algorithm of direction-aware continuous moving *K*-nearest neighbor (DACKNN) queries in road networks. In this method, the objects' azimuth information is adopted to determine the moving direction, ensuring the moving objects in the result set towards the query object. In addition, we evaluate the DACKNN query algorithm via comprehensive tests on the Los Angeles network TIGER/LINE data and compare DACKNN with other existing algorithms. The comparative test results demonstrate that our algorithm can perform the direction-aware CKNN query accurately and efficiently.

Keywords: direction-aware; road network; moving objects; continuous *K* nearest neighbor query

---

1. Introduction

There are many LBS applications, such as taxi hailing, ride sharing and car navigation, and various *K*-nearest-neighbor (KNN) query algorithms have been proposed to solve these problems. With the development of geographic information systems (GISs), KNN query in road networks has evolved from static objects to dynamic objects. The moving objects change their locations and directions frequently over time; therefore, the cost of retrieving the exact results of continuous *K*-nearest neighbor (CKNN) for moving objects is expensive, particularly in highly dynamic spatio-temporal applications, for example, finding the nearest taxi while the user moves in road networks over a period of time. When the user moves to a new location, the traditional solution is to perform a snapshot KNN query. Due to the objects' continuous movement in road networks, a series of frequent snapshot KNN queries are necessary, which are unrealistic and expensive. The cost includes updating the location of moving objects when the velocities change over time and processing the CKNN queries that are posed by the moving object to the server. Therefore, this paper proposes a continuous KNN query method for predicting the *K*-nearest moving objects via predictive computation.

As shown in Figure 1, there are a query object, which is denoted by *q*, and a moving object set, which is denoted by *P*. Each object in the road network moves in the direction of arrow. Taking the query object *q* as an example, *q*(2) indicates that the query object *q* moves at speed 2 towards starting point *n*2 of the edge where *q* is located in the road network. The moving query object *q* requests to retrieve the nearest neighbor object in a specified period. In Figure 1a, the result of the nearest-neighbor query is *p*1 at timestamp 0, and the query result is changed to *p*2 at timestamp 2, as shown in Figure 1b. The query system must monitor the timestamp when the query result will change in the time interval [0, 2] and return the updated result. The query result consists of two-tuples, <{*p*1}, [*t*0, *t*1]>, <{*p*2}, [*t*1, *t*2]> ... <{*pi*}, [*ti*-1, *ti*]>. {*pn*}, [*tn*-1, *tn*]. The query results in Figure 1 are as follows: <{*p*1}, [0, 10/7]> and <{*p*2}, [10/7, 2]>.

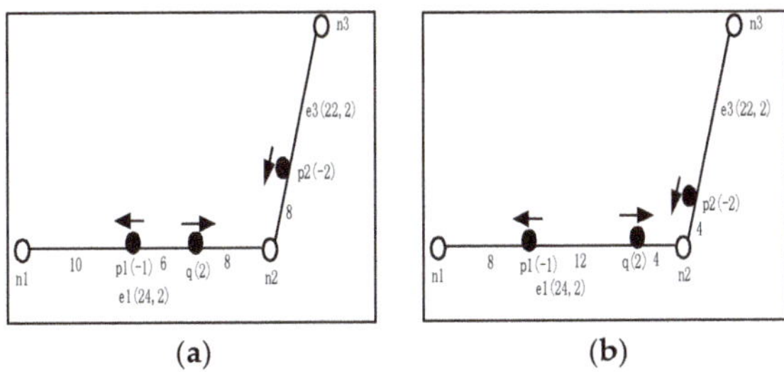

Figure 1. The local road network information under different timestamp: (**a**) at timestamp 0; (**b**) at timestamp 2.

As in Figure 1, the CKNN query only uses distance as the criterion, namely, the KNN query recommended objects are the nearest objects to the query object; however, there may be objects that move away from the query object gradually and the distance to the query object will increase. Moreover, in one-way roads, these moving objects need to turn around at the next intersection to reach the query object. For identifying the object that is nearest to the query object in application scenarios, the object that is queried in Figure 1a is invalid. Therefore, the CKNN query should not only consider the distance factor but also the direction of moving objects. In this paper, we propose a method of direction-aware continuous moving *K*-nearest neighbor (DACKNN) query in road networks. By using the direction constraint, we can continuously monitor the *K*-nearest moving objects that are moving towards the query object.

Many scholars [1–3] have utilized the search-space pruning algorithm, which is based on a road network, to deal with continuous *K*-nearest neighbor queries. Most of them only consider the nearest objects in the dimension of distance. There are some scholars [4–6] that consider the direction-aware query of moving objects in Euclidean space. However, the problem does not involve road network constraints; hence, it cannot be applied to continuous nearest-neighbor query in road networks. In this paper, we consider direction-aware continuous nearest-neighbor queries under the following three assumptions: (1) all objects (including the query object) are dynamic in the road network. (2) The distance between two objects is represented by the shortest path between the two objects in the road network. (3) The result set in each sub-interval of a period is determined. The first problem to be solved is the calculation of the distances between moving objects and the query object at every timestamp in the road network. Because of the continuous movement of objects in the road network, the distance between any two objects is constantly changing. If the Dijkstra algorithm is used for every movement change, the re-computation of the network distance is highly time-consuming. In this paper, the distance between moving objects is calculated based on three factors: the moving speed, the moving direction, and the shortest path between the nodes in the road network. It is a linear function of time.

The main contributions of this paper are as follows:

- The directions of the moving objects are considered to guarantee that the objects in the result set move towards the query object to realize the direction-aware CKNN.
- The concept of monitoring distance is proposed to determine which moving object affects the CKNN query results.
- This paper evaluates the DACKNN algorithm via comprehensive tests on the Los Angeles network TIGER/LINE data and the comparative test results demonstrate the efficiency and practicability of our DACKNN algorithm.

In this paper, Section 2 introduces the related work on the *K*-nearest-neighbor query in road networks; Section 3 presents the data structure, problem definition and related concepts; Section 4 describes the direction-aware continuous moving *K*-nearest-neighbor query algorithm in a road network environment; Section 5 presents the results and analyses of the relevant comparative tests; and Section 6 summarizes the paper, identifies the shortcomings and discusses future work.

2. Related Work

Shifting from static objects to dynamic objects, many scholars focused on CKNN queries under road network constraints. However, the practical requirements of users are not met by only considering the distance factor. Therefore, the direction-based CKNN query will inevitably become a trend in spatial database technology.

2.1. K-Nearest-Neighbor Query in Road Networks

The *K*-nearest-neighbor query in road networks has been widely studied in the field of spatial databases. Papadias et al. [7] proposed a flexible system for handling KNN queries in spatial road network databases, which used the Incremental Euclidean Restriction (IER) algorithm and the Incremental Network Expansion (INE) algorithm to discover KNN objects. These extensions are inspired by the Dijkstra algorithm [8]. Kolahdouzan et al. [9] introduced the Network Nearest Neighbor (VN3) algorithm for solving KNN queries by dividing the spatial network into several small voronoi polygons, and calculated the road network distance based on these voronoi polygons. Hu et al. [10,11] facilitated the KNN query by establishing an index. The KNN query was performed by retrieving a set of interconnected trees [10]. These trees come from the road network. Hu et al. [11] used the distance labeling to maintain approximate road network distances and construct indices to accelerate KNN search. Lee et al. [12] pruned the KNN search space into a subspace without objects. The above studies only consider the distances in the road network as the criterion for judging the neighborhoods; however, in practice, the distance between moving objects cannot be determined only by the physical distance and the relative direction of motion between the two is also an important factor.

2.2. Continuous K-Nearest-Neighbor Query in Road Networks

With the rapid development of moving devices, the motion states of moving objects in space can be monitored in real time. We are no longer satisfied with KNN query in static environments. The KNN query in dynamic environments has attracted more attention and CKNN query has become a research hotspot. The query results of CKNN query processing are computed continuously in the specified period by moving users; thus, the process is highly complicated.

Tao et al. [1] designed a CKNN query processing method for a line edge. This query processing method uses R-tree as the underlying data structure and the KNN query results of a query object on a line can be obtained via a single calculation; however, this method can only query on a pre-defined line. The authors in [2,3] solved the CKNN query for a moving query location on the query path. Cho et al. [2] employed the unique continuous search algorithm which divides the query path into effective intervals by considering the distance between the object and the query object. Within each interval, regardless of where the query moves, the KNN object is the same. Kolahdouzan et al. [3]

introduced the IE/UBA algorithm for identifying CKNN candidates and divided paths into several sub-paths, each of which has the same KNN result set. The objects that are queried in [2,3] are static on the road network; only the query object can move on the path. Since the object that is being queried is moving, Shahabi et al. [13] developed a Road Network Embedding (RNE) technique, which transforms the road network into a higher-dimensional space, retrieves approximate results, and mainly deals with CKNN queries of moving objects on the road network. Mouratidis et al. [14] proposed the incremental monitoring algorithm (IMA) and group monitoring algorithm (GMA) algorithms for updating the results when updates from objects and edges falling in the expansion tree. Demiryurek et al. [15] extended the method for solving the same problem in [14]. The proposed ER-CKNN algorithm avoids blindly expanding the road network when searching for candidate objects and selects candidate objects based on the Euclidean distance between each object and the query object. Huang et al. [16] developed a continuous K-nearest-neighbor query algorithm (CKNN) for moving objects in road networks. This algorithm uses a continuous detection method for moving objects in road networks, eliminates unqualified moving objects and identifies candidate objects, and evaluates candidate objects to determine whether they belong to the K-nearest neighbors of the query objects. Li et al. [17] introduced a continuous KNN (SCKNN) algorithm that is based on a moving state value. This algorithm considers the continuous K-nearest-neighbor queries of query object and the moving objects' moving states, identifies fewer candidate objects, and increases the computational efficiency. Although the existing continuous K-nearest-neighbor query method considers the distance attributes and the motion state of the object, it does not consider the moving direction of the moving object relative to the query object; thus, it cannot accurately realize continuous K-nearest-neighbor query towards the query object.

2.3. Direction-Based KNN Query

In the spatial database query, researchers have introduced direction attributes and carried out some researches on direction-based spatial query technology. To improve the efficiency of direction determination, an open shape-based strategy (OSS) model proposed by Liu et al. [4], which transforms the direction determination between geometric objects and query objects into a spatial topology analysis between open-shape and closed-geometric objects and improves the query efficiency. Patroumpas [5] solved the range query problem of objects that are moving towards query objects in Euclidean space. To quickly determine whether the moving objects are moving towards the query object, a Polar-Tree is generated for each query object via polarization mapping with query objects as poles to efficiently complete the range query and the objects in the result set are moving towards the query object. However, this method can only be applied to scenarios in which the query object is fixed and known and cannot handle random query requests. Nutanong et al. [6] studied visible K-nearest neighbor (VkNN) queries. The main problem is to identify the neighbors that can be seen by the query object; objects that are blocked by obstacles should be excluded. Gao et al. [18] studied the continuous visible K-nearest neighbor (CVkNN) queries, in which the query object can move along a straight edge and the algorithm returns the visible K-nearest object at any point on the edge. The main strategy of CVkNN is to perform only one single-point query for the whole edge instead of one query for each point of the edge. At the same time, an efficient heuristic strategy is used to prune the interest point set and the obstacle set separately, which substantially improves the query efficiency.

Li et al. [19] solved the K-nearest-neighbor query problem in the perspective direction of query object using direction-aware spatial keyword search. Given the location of the query object and the viewing angle range, the algorithm returns the K-nearest neighbors within the viewing angle range. View-field K-nearest-neighbor (VFkNN) query, which was proposed by Yi et al. [20], solves similar problems to direction-aware spatial keyword search; however, and VFkNN supports query processing of moving objects. This method uses a grid to index interest points, accesses the grid within the perspective, and directly excludes the grid outside the perspective, which substantially increases the query efficiency. Because of the high applicability of the grid index to moving objects, the VFkNN algorithm can not only process queries of static interest points but also fully support moving interest points. At the same time, they also solve the problem of continuous VFkNN query. Continuous VFkNN query is divided into two stages: an initial stage and an update stage. In the initial stage, the author implements two query algorithms, namely, naive search and sector search, which are used to process snapshot VFkNN queries. In the update stage, the sector surveillance algorithm, which is proposed in this paper, can efficiently handle the updating of moving objects. Lee et al. [21–23] studied the nearest surrounder (NS) query. No longer is only the interest point that is nearest to the query object identified; multiple interest points are returned. These interest points are centered on the query object, distributed in various directions of query object and have a unique interest point in each direction. These interest points become the surroundings of the query object, that is to say, NS gives the query object a global perspective, from which the nearest-neighbor object to the query object in any direction can be identified. Guo [24] and others proposed Direction-Based Surrounder (DBS) queries for solving NS queries in Euclidean space and road network space and supporting continuous DBS query processing. Lee et al. [25] solved the K-nearest-neighbor query problem with the same direction as the query object by using a direction-constrained K-nearest-neighbor (DCkNN) query. Given the location and direction of query object, DCkNN can identify a point in the direction of the query object from the set of interest points if the difference value does not exceed a specified threshold, which is denoted as θ. Chen et al. [26] determined the nearest-neighbor query of the path and retrieved the nearest-neighbor object along the path of the query object.

Although these KNN query methods consider the direction attribute, the solved problems do not involve road network constraints or continuous moving K-nearest-neighbor query. Therefore, we propose a direction-aware continuous moving K-nearest-neighbor query algorithm in road networks that is based on progressive network expansion. Our previous research has solved the problem of Direction-aware KNN queries for moving objects in a road network (DAKNN) [27]. In this paper, we focused on the problem of continuous queries and the efficiency of continuous K-nearest-neighbor query for moving objects in road networks that are moving towards query objects.

3. Data Structure and Problem Definition

In this section, we present important concepts and data structures that are involved in directional-aware continuous moving K-nearest neighbor queries and formally define the problem that we attempt to solve in the paper.

3.1. Data Structure

In our method, it is assumed that the speed of each object that is moving in road networks is constant. The data structures are designed as follows.

The road network is composed of nodes and edges, which is represented by an undirected weighted graph, namely, G (N, E, W), where N represents the set of nodes and stores all node information of the road network; E represents the set of edges and stores the edges of the road network; W represents the weight of the edge, where the weight that is set by the system is the length of the edge [27]. In addition, the two endpoints of a specified edge are denoted by ns and ne. ns represents the starting point and ne represents the ending point of the edge. For simplicity, we do not consider the direction of each edge; therefore, the starting points and ending points of the edges in the road network are represented by only two endpoints, which are arbitrarily specified.

The moving objects are a set $S(t)$, where $p(t) \in S(t)$ is a moving object p in the road network at timestamp t. The spatial information for each moving object $p(t)$ in the road network, such as its location, speed, and movement direction at a timestamp, is available. Especially, when the moving object reaches a road network node, an adjacent edge of the node is randomly selected as its next traveling edge, the moving object will travel along the selected edge.

The query object refers to the spatial object that makes the query request, such as when a pedestrian makes the following query request: "At a certain period of time, find two taxis coming toward me." In this case, the pedestrian is the initiator of the query request, namely, the query object. The query object typically contains the following information: (1) the query coordinates; (2) the query time interval; (3) the value of K; and (4) the result set.

Given a range value m, starting from the query object, the node is gradually expanded until the distance from all adjacent nodes of the current node to query object is greater than the range value m. Expanded edges constitute the influencing edge, which is composed of triples $(ni, nj, dist)$, where $dist$ represents the distance between the nj node and ni.

In our algorithm, a data structure of memory is used to store edge, node and moving object information [27].

- Edge table: the following information is stored: (1) the edge starting and ending nodes; (2) the lengths of the edges; (3) the sets of objects that are moving on the edges; (4) the directions of the edges; and (5) the maximum speed.
- Node table: this mainly includes the node identifiers and the location information.
- Moving object table: the following information is stored: (1) the edge of the object; (2) the moving speed; (3) the distance to the starting node; (4) whether it moves towards the ending point; (5) the moving direction of the object; and (6) the time when the location of the object is updated.

3.2. Road Network Distance

The distance between two objects on an undirected weighted graph is the length of the shortest path. If two objects are not reachable, the distance is ∞.

In this paper, we must determine the road network distances between moving objects at each timestamp in the time interval. The distance between the moving object p and the query object q is denoted as $ND_{q,p(t)}$. When the object arrives at the network node, the network distance $ND_{q,p(t)}$ must be recalculated. If the query object q and the moving object p are specified, the distance between them in any time interval can be calculated. According to the edge of the moving object and the location of the query object, there are two cases.

In the first case, the query object and the moving object move on the same edge, namely e, and the distance $ND_{q,p}$ is calculated as follows:

$$ND_{q,p(t)} = \left|(q.distTos + q.signS * (t - q.tu)) - (p.distTos + p.signS * (t - p.tu))\right| \qquad (1)$$

In an undirected weighted graph, the starting and ending nodes are arbitrarily specified on each edge. If the object moves towards the ending node, $signS$ indicates the positive speed; otherwise, $signS$ indicates the negative speed. $q.signS$ and $q.tu$ represent the marking speed and the last update time, respectively, of the query object. Similarly, $p.signS$ and $p.tu$ represent the marking speed and the last update time, respectively, of the moving object. $q.distTos$ represents the distance between the query object and the starting point of the edge that is located by the query object. When an object arrives at the road network node, the distance between the object and the query object must be recalculated.

In the second case, query object q and the moving object p move on different edges, e.g., edge $ei(nis, nie)$ and edge $ej(njs, nje)$. For simplicity, in this paper, we pre-calculate the shortest distance between the nodes in the road network and use $SP(ni, nj)$ to represent the shortest path between nodes ni and nj. In this case, the distance between objects can be calculated by the shortest path between nodes. The shortest paths between the starting nodes and the ending nodes of the two moving objects are denoted as $SP(nis, njs)$, $SP(nis, nje)$, $SP(nie, njs)$, and $SP(nie, nje)$. Therefore, the distance between the moving object and the query object at timestamp t is composed of the following three parts: (1) the distance between the query object and node nis or nie; (2) the shortest path between the nodes; and (3) the distance between the moving object and node njs or nje. The road network distance can be expressed as:

$$ND_{q.p(t)} = \min\left\{ND1_{(t)}, ND2_{(t)}, ND3_{(t)}, ND4_{(t)}\right\} \tag{2}$$

$$ND1_{(t)} = q.distTos + q.signS * (t - t.tu) + SP(nis, njs) + p.distTos + p.signS * (t - t.tu)$$
$$ND2_{(t)} = q.distTos + q.signS * (t - t.tu) + SP(nis, nje) + p.distToe - p.signS * (t - t.tu)$$
$$ND3_{(t)} = q.distToe - q.signS * (t - t.tu) + SP(nie, njs) + p.distTos + p.signS * (t - t.tu)$$
$$ND4_{(t)} = q.distToe - q.signS * (t - t.tu) + SP(nis, njs) + p.distToe - p.signS * (t - t.tu)$$

where $p.distTos$ represents the distance between the object and the starting point of the edge and $q.distToe$ represents the distance between the object and the ending point of the edge. For example, in Figure 2, the paths are composed of edges $e1(n1, n2)$ and $e3(n2, n3)$. The moving object, namely, $p3$, and the query object, namely, q, are located at $e3(n2, n3)$ and $e1(n1, n2)$, respectively. The object $p3$ moves towards the starting point, namely, $n2$; hence, $p3.signS$ is -1. The query object q moves towards the ending point, namely, $n2$; hence, $q.signS$ is 1. The distance between them is $ND_{q.p3(t)} = 14 - 2t$. We can find that when the moving object arrives at the node, its moving edge will change, and the four nodes involved in the distance calculation formula also change; thus, once the moving object reaches the node, its distance formula to the query point needs to be recalculated.

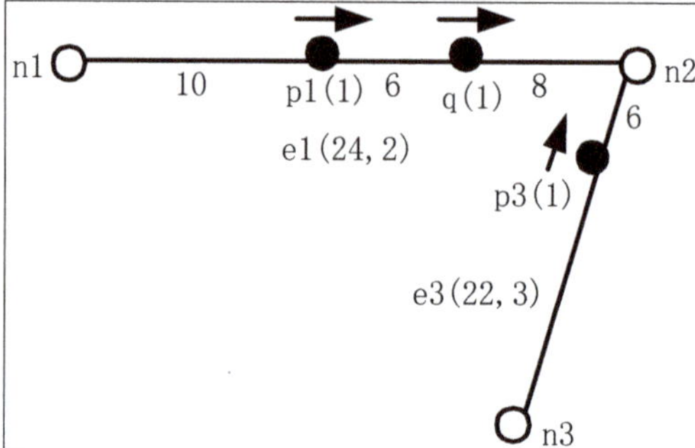

Figure 2. The local road network information.

3.3. Problem Definition

After the above overview and basic definitions, we define the problem that is considered in this paper.

Question Definition: In this paper, we study direction-aware continuous moving KNN queries in road networks in dynamic environments.

In a dynamic environment, if there is a group of moving objects P and a query object q in the road network, the query is to retrieve the K-nearest neighbors from the road network at any timestamp in $[t0, t1]$ towards the query object, e.g., "A moving passenger inquires about K taxis moving towards himself during $[t0, t1]$ period".

For the example in Figure 3, the global problem is clarified. In edge $ei(w, s)$, w represents the length of the edge and s represents the speed limit of ei indicating the maximum speed of the moving object on ei. For moving object $p(s)$, s represents the speed of object p. The query point q requests the two nearest-neighbor objects in the time interval $[0, 4]$.

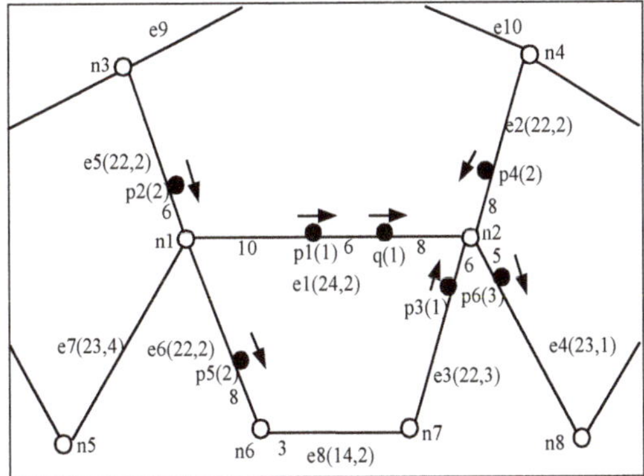

Figure 3. Road network at timestamp 0.

4. Direction-Aware Continuous Moving *K*-Nearest Neighbor Query

When the query object arrives at the road network node, it records the timestamp and divides the time interval into sub-intervals such that in each sub-interval, the query object moves on the same edge. First, we must execute the direction-aware KNN query algorithm at the starting timestamp to identify the *K*-nearest neighbor objects that are moving towards the query object. Then, the monitoring range of continuous queries is calculated; the moving objects in the monitoring range may affect the query results. After that, a local road network is established according to the monitoring range and then moving candidates are identified. Finally, the timestamp when the candidate objects replace the query results is determined. If the timestamp is within the specified period, the sub-interval is divided, and the result set is modified. Figure 4 shows the basic process of the algorithm.

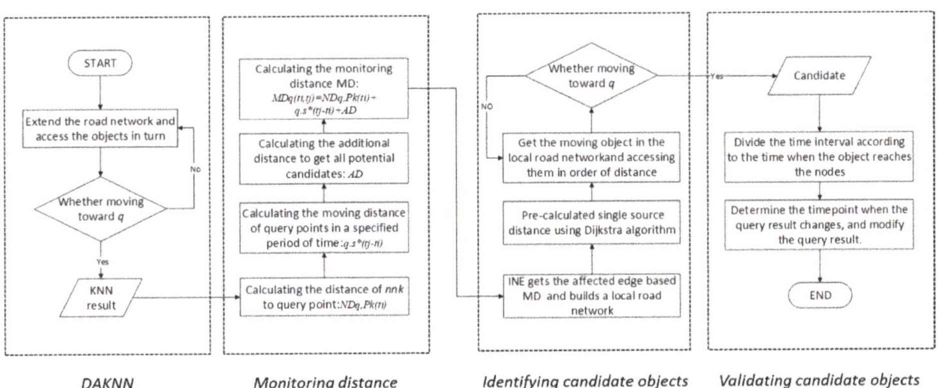

Figure 4. Direction-aware continuous moving *K*-nearest neighbor (DACKNN) algorithm process diagram.

4.1. Direction-Aware K-Nearest Neighbor Query for Moving Objects

In determining whether the moving object moves towards query object in road networks, two cases are considered. Case 1: the moving object and the query object are on the same edge. At this time, it is only necessary to determine whether the moving direction of the moving object points towards the query object or not. The moving direction of the moving object is determined according to the azimuth information. Case 2: the moving object and the query object are on the different edge. This paper adopts the method of Direction-aware K-Nearest Neighbor Query [27], in which road network expansion and azimuth angle are used to determine k neighbor objects moving toward the query point. The azimuth angle is used to determine if the direction of moving object is toward the query point, and the direction determination rule is the direction of road network expansion is opposite the direction towards the query object.

As shown in Figure 5, the horizontal angle between the reference point, which is denoted as $n1$, and the target directional line, which is denoted as $n1n2$, is α. In the road network, node ni is either the starting node or the ending node of edge; hence, the azimuth angle, namely, β, of the road network expansion is $\beta = \alpha$ or $\beta = \alpha + 180\,mod\,360$. In Figure 6, if the current accessing node is $n3$, its extended adjacent edges are $e3$ and $e4$, the extended direction has two, when the extended edge $e3(n3, n6)$, and the extended direction is $n3$ to $n6$, then $\beta = \beta3$; when the extended edge $e4(n3, n4)$ and the extended direction is $n3$ to $n4$, $\beta = \beta2$.

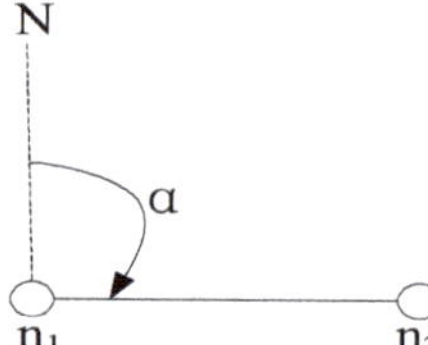

Figure 5. Azimuth angle.

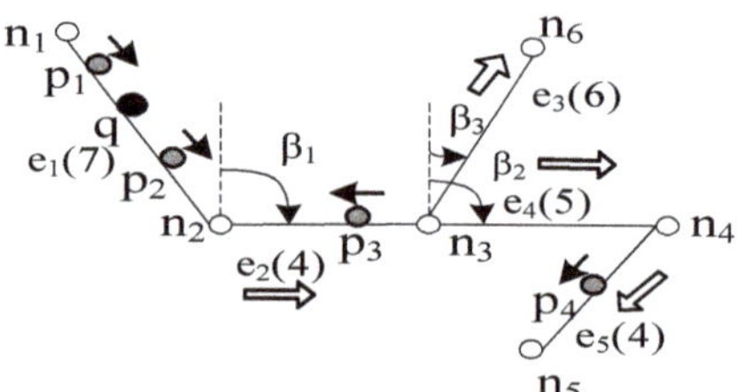

Figure 6. Azimuth angle of road network expansion.

Using the direction of road network expansion, the rules for judging whether the moving object is moving on road network towards query object are as follows: when the road network extends to the adjacent edge, which is denoted as $ei(ni, nj)$, the direction of moving object is denoted as γ, the direction of road network expansion is denoted as β. The moving object is moving towards the query object if it satisfies the inequality $\varepsilon\text{-}180 \leq \gamma\text{-}\beta \leq 180\text{-}\varepsilon$ (ε is the minimum deviation of the azimuth).

Based on the method for identifying the moving objects of road network that are moving towards query object, this paper uses R-tree to quickly retrieve the road edges and an INE-based road network expansion method to query the K-nearest moving objects, which efficiently identifies the K-nearest objects that are moving towards the query object. The algorithm generates a priority queue according to the distance from each network node to the query object and the priority queue controls the expansion order of the network. The closer a node is, the earlier its adjacent edges will be expanded. If an adjacent edge e of node v is currently expanding and all the moving objects on the adjacent edge have been retrieved, then use the azimuth angle to judge whether these objects are moving towards the current road network node v. If p is moving towards v, the road network distance from p to the query object is calculated, and put p into the result set. Then, the adjacent edge e is marked as "expanded" and this process will continue until there are K objects in the result set. Only when the distance from the next road network node to the query object is smaller than the distance from the farthest neighbor of the result set to the query object; otherwise, the algorithm directly returns the K objects that are moving towards the query object.

For the problem in Figure 3, since the speed of q is 1, the time to reach road network node $n2$ is 8; therefore, in the time interval $[0, 4]$, q always moves on the same road edge. In this stage, we must identify the K-nearest neighbor object that is moving towards the query object at timestamp 0.

Combined with the INE road network extension and the direction determination method, the algorithm process is as follows: first, edge $e1$ is retrieved, object $p1$ is acquired, $p1$ is moved towards the query object, and the distance between $p1$ and the query object is inserted into the result set: $R = \{(p1,6)\}$. Node $n2$ is closest to the query object and adjacent edges $e2$, $e3$, and $e4$ of node $n2$ are extended, $p4$ is retrieved on edge $e2$ and the direction of $p4$ is opposite the direction of expansion. Thus, $R = \{(p1, 6), (p4, 16)\}$. Next, $p3$ at edge $e3$ is retrieved, since the distance from the query object is 14, which is smaller than p4, $R = \{(p1, 6), (p3, 14)\}$ is inserted; on edge $e4$, $p6$ is retrieved and it is determined that $p6$ is consistent with the expansion direction; the moving direction is far from the query object, and the condition is not satisfied. Then, the other end node, namely, $n1$ of $e1$, is expanded and the moving object on the adjacent edge of $n1$ is retrieved. The algorithm terminates when the distance of the next node of the algorithm is larger than the distance of the second-nearest-neighbor object to the query object. The final result is $R = \{(p1, 6), (p3, 14)\}$.

4.2. Evaluation of Monitoring Range in Continuous Queries

The objective of this stage is to obtain the road network monitoring range of the query request. For the time sub-interval, namely $[ti, tj]$, of the query, the direction-aware road network moving object KNN query technology is used to identify the KNN object that is moving towards the query point and the result set is $\{p1, p2 \ldots, NNk\}$. In this result set, NNk is the object farthest from the query point. The monitoring distance (MD) is calculated according to the result set, and it ensures that only an object that is within the monitoring distance can become the final continuous query result. Knowing that the objects in the KNN result set are all moving towards the query point and the distance between NNk and the query point is the largest at timestamp ti. The MD is calculated as follows:

$$MD_{q(ti,tj)} = ND_{q,NNk(ti)} + q.s * (tj - ti) + AD \tag{3}$$

There are three parts as follows: (1) $ND_{q,NNk(ti)}$ represents the distance between NNk and the query point at timestamp ti; (2) $q.s * (tj - ti)$ represents the distance that query point q moves between $[ti, tj]$; and (3) AD represents an additional distance to ensure that all potential candidates for the KNN results are monitored in the query process within the time interval $[ti, tj]$. AD is calculated as follows: firstly, the set of edges arrived after extending the distance of $ND_{q,NNk(ti)} + q.s * (tj - ti)$ from the query point is determined and the additional distances of these edges are calculated as $ei.s * (tj - ti)$. Then, the maximum additional distance is $AD = \max\{ei.s * (tj - ti)\}$.

If the distance from an object on an edge to a query point exceeds $ND_{q,NNk(ti)} + q.s * (tj - ti)$, but this object moves towards the query point at the maximum speed for this edge, the moving object may be the result of the continuous KNN query. Therefore, we must add an additional distance to avoid missing all possible KNN results.

We consider the example in the previous section. After the DAKNN query processing stage, we obtain K-nearest neighbor object R = $\{(p1, 6), (p3, 14)\}$, $ND_{q,p1(0)} = 6$, $ND_{q,p3(0)} = 14$. The distance that q moves in time interval [0, 4] is 4 and the road network monitoring distance is MD = $14 + 4 + AD$, as shown in Figure 7. The objects $p1$ and $p3$ in the result set move towards the query point; hence, the distance between $p1$ or $p3$ and query point must be less than 18 in the time interval [0, 4]. That is, at timestamp 4, the query point arrives at q' and a distance of 18 is extended from the query point q' and marked with dotted lines. In Figure 7, the marks fall on edges $e1$, $e2$, $e3$, and $e4$ and AD = max $\{2*4, 2*4, 3*4, 1*4\} = 12$. Therefore, the monitoring range of the road network is 30 and by expanding by a distance of 30 from query point q', the monitoring range is obtained and marked with solid lines in Figure 7.

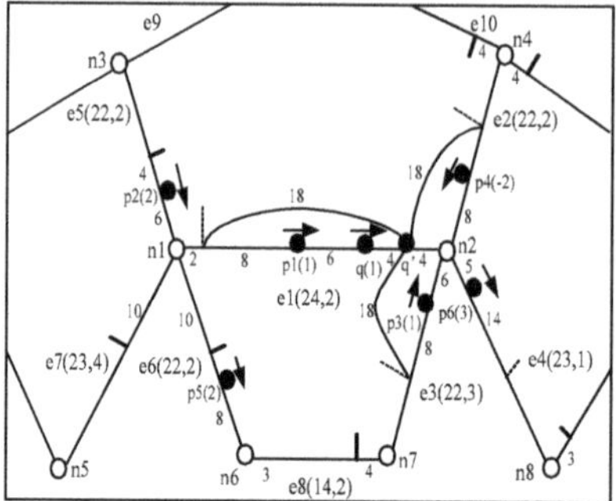

Figure 7. An example of monitoring range.

4.3. Identifying Candidate Objects

This stage is a process to prune the search space and gets all moving objects for which the road network distances are less than MD. The task at this stage is mainly to build a local road network and to obtain candidate objects.

The first task is to build a local road network; all the edges in the monitoring range are added to the list of influential edges for constructing a local road network. We use the road network expansion method to expand the road network from near and far from the road edge where the query point is located. For one road edge, if the distance from one of the nodes to the query point is less than MD, add it to influence edges to participate in the construction of the local road network. It is specifically stated that the Dijkstra algorithm is used to pre-calculate the shortest distance from the nodes in the local road network to the query point. The purpose of pre-calculation is to avoid inefficiencies caused by online distance calculation. That is to say, in the subsequent calculation process of the distance from the moving point to the query point, the pre-calculation improves the efficiency of the algorithm. The information in the local road network is useful information that may affect the query result and the moving object in the local road network is also the candidate object of the query result. Our next task is to obtain the moving object moving towards the query point in the local road network as the candidate object, then use the proposed method to determine the direction of moving object. As shown in Figure 8, the final set of candidate objects is {$p1, p2, p3, p4$}.

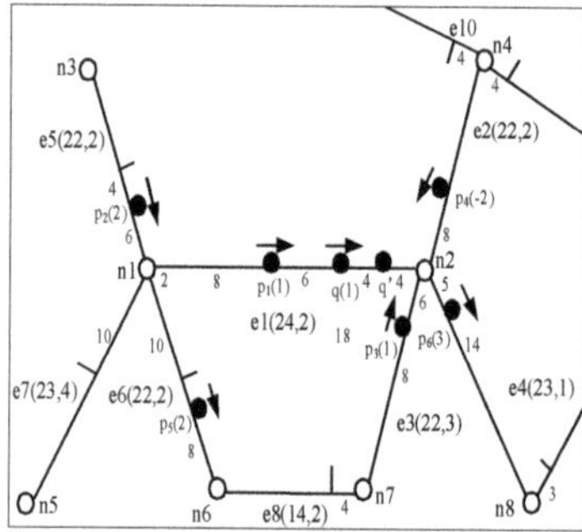

Figure 8. The Local Road Network.

4.4. Validating Candidate Objects

After the pruning stage, any object that cannot be the query result is excluded and a candidate object set, namely, {$p1, p2 \ldots, pm$} ($m \geq k$), in the specified interval, namely [$t0, tn$], is obtained. Since the object is continuously moving, the query result may change within this time interval; if we cannot determine the time point at which the change occurs, queries at multiple timestamps maybe return the same query result. To reduce the cost of repeated query, we proposed the verification algorithm for determining the time points at which the time interval is divided into sub-intervals, which ensures that the query result set remains unchanged over each sub-interval.

In the verification algorithm, we first divided the specified time interval, namely, $[t0, tn]$ into several sub-intervals, namely, $[t0, t1], [t1, t2] \ldots [tm - 1, tm], [tm, tn]$, based on the time points when the candidates arrive at the nodes of the road network. When the moving objects arrive at the network nodes, the formula for calculating the distance between them and the query point will change. For each sub-interval, the distance between the moving object and the query point is determined via a specified linear function of time.

The objective of the verification algorithm is to identify the time point tc of the KNN result set change in each sub-interval $[tm - 1, tm]$, so that the result set is the same at any timestamp within two consecutive tcs. Using the DAKNN method, which was introduced in Section 4.1, to obtain the query result set Q of initial time $tm - 1$, the first object to be replaced by other candidate objects in the query result set Q must be NNk. The distance between the moving object NNk and the query point should be recorded as $ND_{q,NNK(t)}$. For each of the candidate objects, the time point at which NNk is replaced by pi can be calculated via equation $ND_{q,NNk(t)} = ND_{q,pi(t)} (tm - 1 < t < tm)$ and the minimum calculated time point can be selected as the time point of change $tc1$. There are two cases as follows: (1) if object pi is in the KNN result set, the query result only changes in order and pi becomes a new NNk; (2) If object pi is not in the KNN result set, time subinterval $[tm - 1, tm]$ is divided into $[tm - 1, tc1]$ and $[tc1, tm]$. For partitioned $[tm - 1, tc1]$, the query results at any time in the subinterval are consistent with those at time $tm - 1$. For partitioned $[tc1, tm]$, NNk of query result Q is replaced by pi as the query result at time $tc1$ and the above calculation process is repeated until the time of change exceeds tm.

We proceed with the example from the previous stage to verify that we have obtained the result set of candidate objects: $\{p1, p2, p3, p4\}$. In Figure 8, we divide the time interval, namely, $[0, 4]$, into sub-intervals according to the time when the $p2$ arrives at the road network node: $[0, 3]$ and $[3, 4]$. First, we verify subinterval $[0, 3]$. At time 0, KNNs = $\{p1, p3\}$ and $NNk = p3$. Using the equation $ND_{q,NNk(t)} = ND_{q,p(t)}$ to verify each candidate, it is concluded that object $p4$ replaces $p3$ at time 2 and $p4$ is not in the KNN result set; thus, the result set changes at time $tc = 2$. Therefore, time sub-interval $[0, 3]$ is divided into $[0, 2]$ and $[2, 3]$ and the query results are <$\{p1, p3\}, [0, 2]$> and <$\{p1, p4\}, [2, 3]$>. For time interval $[3, 4]$, the candidates are validated via the same processes. At time 10/3, $p1$ replaces $p4$ and $p1$ is in the KNN set; the final result is R= <$\{p1, p3\}, [0, 2]$>, <$\{p1, p4\}, [2, 10/3]$>, and <$\{p4, p1\}, [10/3, 4]$>.

5. Results and Analysis

This section evaluates the DACKNN query algorithm in road networks via detailed comparative test schemes. The performance of the algorithm is measured in terms of the execution time of the CPU. The results demonstrate that the proposed DACKNN method can efficiently process direction-aware continuous K-nearest neighbor queries of moving objects in road networks.

5.1. Parameter Setting

The comparative test is implemented in Java and run on an Intel (R) Core (TM) i5-6200U CPU @ 2.30 GHz processor, 4G memory, and 64-bit Windows 10 operating system. The comparative test data in this paper include road network data and moving object data, as presented in Table 1. The road network data are based on Los Angeles from TIGER/Line [28] and include 195,888 road nodes and 267,536 edges. At the same time, the Brinkoff Road Network Data Simulator [29] was used to generate moving object data with moving directions. When the moving object arrives at a road network node, it randomly selects the next path. These data objects are evenly distributed in the road network, and there are 10,000–100,000 of them.

Table 1. Comparative test data.

Parameters	Values
Los Angeles Edge	267,536
Los Angeles Node	195,888
Moving object	10K~100K
K	1~100
Time interval t	0~100

The performance of the comparative test algorithm is mainly measured in terms of the CPU execution time. The effects of the number of moving objects, the number of K-nearest neighbors and the time interval of continuous query on the query performance are investigated via the control variable method. Because the location of the query point affects the execution time, we execute 1000 non-concurrent queries at various locations and the average execution time is taken as the comparative test result for analysis. The default settings of each parameter are adopted by the current mainstream continuous nearest neighbor queries of moving objects that are based on the road network, as listed in Table 2.

Table 2. Default parameter settings for comparative tests.

Parameters	Default Values
R-tree maximum capacity of nodes	50
R-tree minimum capacity of nodes	20
Moving object p	50k
K	30
Time interval t	30

In addition, we analyzed the performances and accuracies of the CKNN, SCKNN, and DACKNN algorithms, and demonstrated the advantage of DACKNN.

5.2. Comparative Test Results and Analysis

First, the DACKNN algorithm is analyzed under various parameter settings. Second, the time efficiency and accuracy are compared with those of the SCKNN and CKNN query algorithms. Finally, the shortcomings are identified to facilitate improvement in future work. In this comparative test, default values are used for all variables unless otherwise specified.

5.2.1. Evaluation of the DACKNN Algorithm

We conduct a series of comparative tests in which we vary the number of neighbors: K, the number of moving objects and the time interval and analyze the performance of DACKNN under the parameter settings.

(1) The influence of the number of moving objects on the query performance. According to Figure 10, the value of K is 30. In the comparative tests, we compared the effects of the number of moving objects on the overall continuous query performance in three-time intervals: [0, 30], [0, 60], and [0, 90]. As shown in this Figure 10, the overall query time increased with the number of moving objects and when the number of moving objects is between 10,000 and 20,000, the query performance among the query intervals tended to be stable. This is because the number of objects in the local road network that was established by the monitoring range increased and the time cost of judging whether each moving object was moving towards the query point and executing the verification algorithm also increased. As shown in Figure 14, according to the number of moving objects in the global road network, the number of moving objects in the local road network that was established by the monitoring range of time intervals [0, 30], [0, 60], and [0, 90] is displayed. The figure shows that as the number of moving objects in the global road network increases

gradually, the number of objects in the local road network also increases. The more objects there are, the longer it takes to evaluate the candidate objects. Hence, the complete algorithm becomes more time-consuming as the number of moving objects increases. From Figure 10, we also can see the larger the time interval, the higher the time cost of the algorithm. In different time intervals, when the interval difference is not very large, the calculated monitoring distances are similar and the numbers of moving objects in the local road networks that were established according to the monitoring ranges are almost the same. However, the longer the time interval is, the more time-consuming the whole algorithm will be, because the longer the time interval is, the number of moving objects that have arrived at the nodes increases, the time interval is divided into several sub-intervals according to their arrival times, and the road network distances between them and the query object are recalculated; thus, the time-consuming will increase accordingly.

(2) The influence of the number of neighbors on the performance. Figure 9 shows the query execution in the Los Angles road network with 50,000 moving objects. The overall query time varies with the K under the query intervals: [0, 30], [0, 60], and [0, 90]. As shown in Figure 9, when the numbers of moving objects are the same, as K increases, the overall query running times will increase. This is because the scale of local road network gradually increases with an increase of K, thereby resulting in an increasing number of objects to be verified. Figure 13 shows how the numbers of edges, nodes and moving objects in a local road network vary with the K value. With the increase of K, the scale of local road network increases gradually and the number of moving objects in the monitoring range of the local road network also increases gradually. If the number of neighbors, namely, K, is 10, the number of moving objects in the monitoring range is 823 and the numbers of edges and nodes in the local road network are 438 and 361, respectively. When the number of neighbors, namely, K, is 30, the number of moving objects in the monitoring range is 1021 and the numbers of edges and nodes in the local road network are 812 and 641. The K value of the nearest neighbors of the query requests increases slightly and the number of local road networks increases gradually. Thus, as the scale of the road network increases steadily, the amount of data that the query algorithm must process and verify increases and the overall query time cost also increases. From Figure 9, we can also see that the larger the time interval, the higher the time cost of the algorithm. Consistent with the analysis in the previous section, the longer the time interval is, the more sub-intervals must be updated and verified.

(3) The influence of the time interval on the overall performance. For 40,000, 50,000, and 60,000 moving objects and 30 nearest neighbors, comparative tests were carried out in various time intervals. The results are shown in Figure 11. As the length of the query time interval increases, the running time increases gradually. The larger the time interval is, the larger the monitoring range is and the larger the number of moving objects that were in the monitoring range. Moreover, with a longer time interval, more candidates arrive at the nodes of the road network in the continuous monitoring range. It is necessary to re-select the moving section, update the calculation formula for the distance from the query object, and divide the sub-intervals. These factors will lead to a higher query time cost as the time interval increases. When the moving objects are distributed in the road network with 265,536 edges, the graph shows that when the number of moving objects is small, the overall query time consumption is stable and the overall query time consumption gradually increases with the time interval. The more moving objects there are in the same time interval, the more moving objects must be retrieved and validated and the longer it takes.

(4) The time-consuming situation of each stage in the algorithm. First, the comparative test makes the following comparisons according to the number of neighbors in the query: when the number of moving objects is 30,000 and the query time interval is [0, 30], the query time distribution for the K-nearest neighbors is adopted. In Figure 12, the histogram represents the time distribution of each stage of the algorithm. The first stage of the algorithm identifies the K-nearest neighbor objects that are moving towards the query point. This stage consumes a small proportion

of the total time consumption of the algorithm. In the process of continuous monitoring of nearest-neighbor queries, the monitoring range is calculated by traversing the KNN result set, which costs little time. The time consumption of the stage of identifying candidate objects via road network expansion accounts for the majority of the total time consumption of the process. In this stage, the local road network is constructed according to the monitoring range and the moving objects in the local road network are evaluated. In the process of identifying the moving objects, according to the diagram, the time-consumption occupancy ratio is the highest out of the whole process when the number of *K*-nearest neighbors is small. In the process of validating the candidate objects, the time interval is divided into several sub-intervals according to the arrival timestamps of the candidate objects at the node of the road network and a result set is obtained in each sub-interval. When dividing the time interval, it is necessary to update each object that arrives at the node and to recalculate the road network distance to the query point. The more candidate objects and *K*-nearest neighbors, the higher the time consumption of the validation process; the time consumption of the verification stage will also increase.

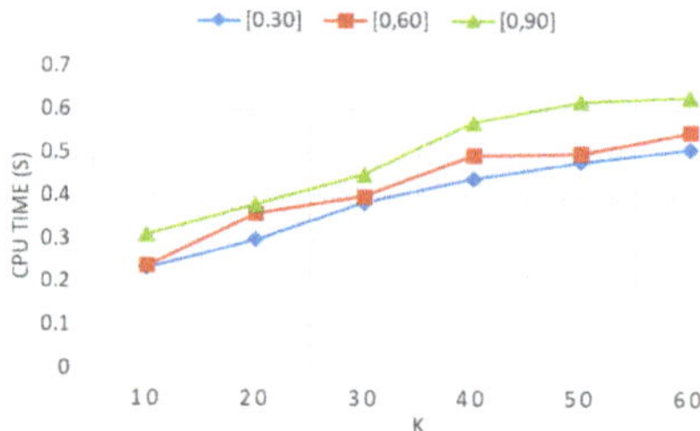

Figure 9. Effect of K value on Performance.

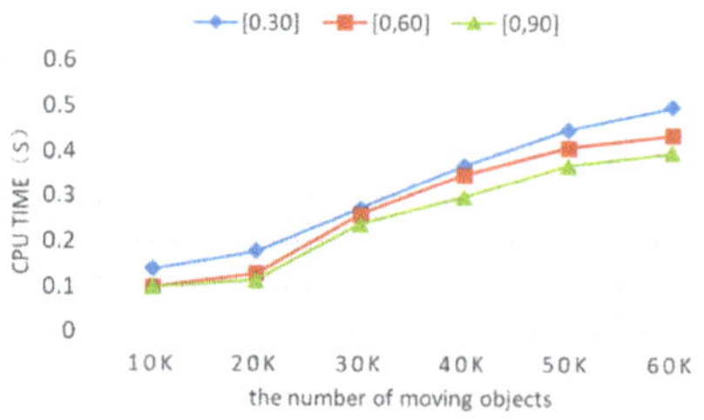

Figure 10. Comparisons of time consumed with the number of moving objects in different time intervals.

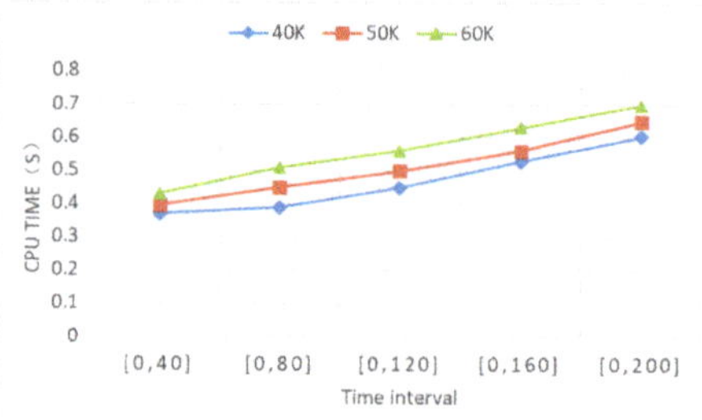

Figure 11. Effect of different time intervals on performance.

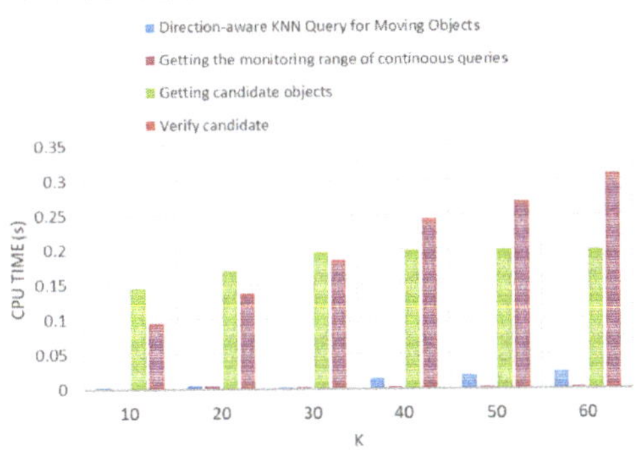

Figure 12. Time-consuming for each stage of continuous query for different K values.

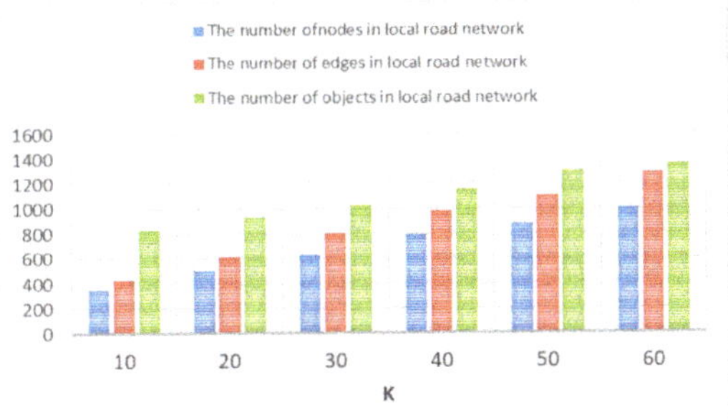

Figure 13. The number of nodes, edges, and moving objects in local road network with different K values.

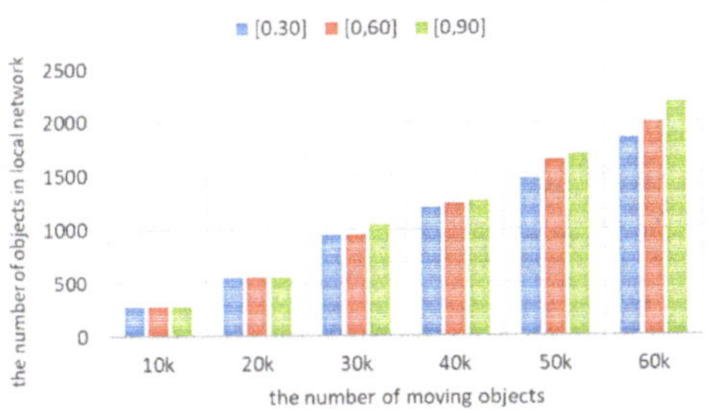

Figure 14. The number of objects in local road networks varies with the number of moving objects in different time intervals.

Then, the comparative test makes the following comparisons according to the various query time intervals. When the number of moving objects is 30,000 and the number of neighbors, namely, K, is 10, the time distribution of each stage of the algorithm is observed for various query time intervals: [0, 20], [0, 40], [0, 60], [0, 80], and [0, 100]. In Figure 15, the histogram shows the time distribution of each stage of the algorithm and the time-consumption trend is stable for acquiring direction-aware neighbors and monitoring ranges. When the number of K-nearest neighbors is constant, the time-consumptions of the K-nearest neighbor queries that are acquired at the starting time are stable when the starting times are the same. In the whole process, it is time-consuming to obtain and verify candidate objects. In the stage of obtaining candidate objects, the influential edges are identified according to the monitoring range, a local road network is created, and the moving objects in the local road network are retrieved as candidate objects. With the increase of the time interval, the monitoring range becomes larger and the amount of data that must be processed increases. The time-consumption of the candidate validation stage increases with the time interval. The longer the time interval is, the larger the monitoring range is, the more edges that must be retrieved, and the larger the number of candidates that must be evaluated. Moreover, the longer the time interval is, the larger the number sub-intervals into which it must be divided. For each sub-interval, the candidates must be evaluated; thus, the time-consumption increases with the number of sub-intervals.

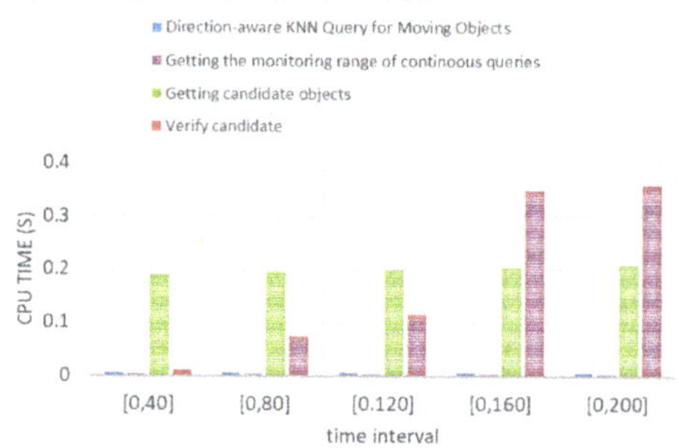

Figure 15. Time-consuming for each stage of continuous query in different time intervals.

To deal with the continuous *K*-nearest-neighbor query and improve the query efficiency, we must transform the global network into a small-scale local network to facilitate the retrieval of objects. We analyze the time-consumption of the whole process of constructing the local network. As shown in Figure 16, in the process of local road network construction, the time consumption of determining the monitoring range is very small. According to the size of the monitoring range, the query point-centered edge is acquired and the local road network is constructed on the edge of the monitoring range; this process is time-consuming. As the time interval increases, the monitoring range increases and the acquisition influence edge becomes larger, thereby increasing the processing time. The time consumption for calculating the shortest path of a single source on a local road network is negligible.

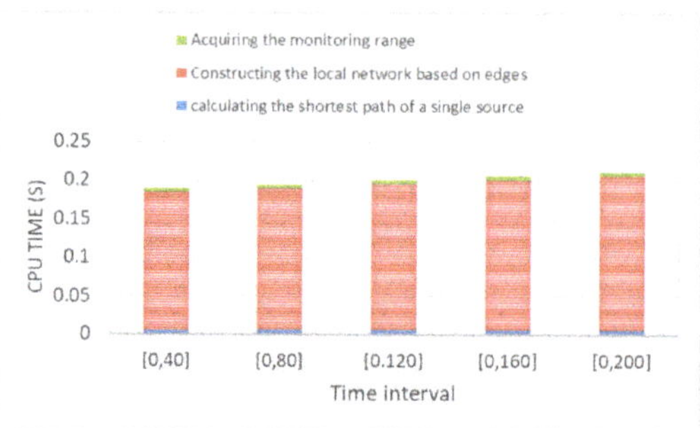

Figure 16. Time-consuming of local road network construction stage under different time interval.

5.2.2. Comparison with Other Algorithms

First, we evaluate the effects of the CKNN, SCKNN, and DACKNN algorithms on the time consumption as the time interval increases and their accuracies. Here, accuracy refers to the accuracy with which moving objects that are moving towards the query point are identified, which is expressed as a percentage. According to Figure 17, the time costs of these three algorithms increase with the time interval because more moving objects will arrive at the network nodes over a larger time interval; thus, the time interval is divided into more sub-time intervals for verification. However, the time costs of CKNN and SKNN increase faster as the time interval increases because these two algorithms do not screen out objects that are moving towards the query object, but consider all objects in the monitoring range as candidates; thus, the more objects they deal with, the more time they consume. The DACKNN algorithm can filter out objects that are moving away from the query point in the monitoring range. Thus, the number of candidate objects is smaller; therefore, DACKNN outperforms SCKNN and CKNN in terms of time consumption. Figure 18 shows the accuracies of these three algorithms. The moving objects that are moving towards the query object can be obtained with 100% accuracy by the DACKNN algorithm as the time interval increases, while the accuracy of SCKNN reaches more than 70%, and the accuracy of CKNN is approximately 50%.

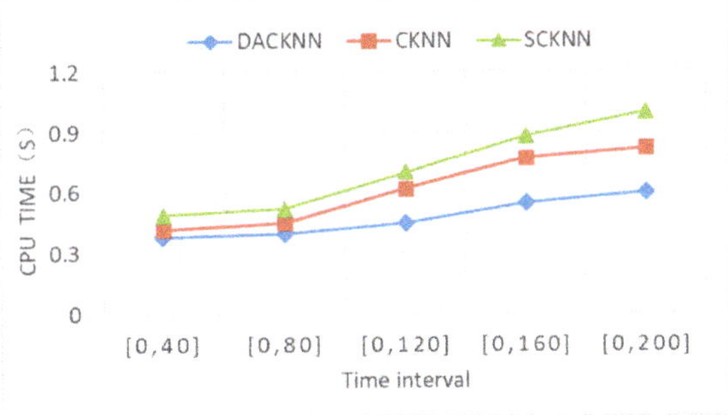

Figure 17. Time consumption of algorithms under different time intervals.

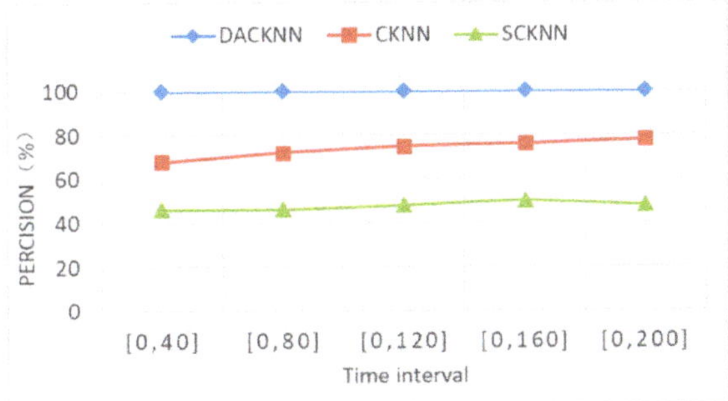

Figure 18. Accuracy of algorithms under different time intervals.

Figure 19 shows that the time costs of CKNN, SCKNN and DACKNN will increase as the value of K increases. This is because with the increase of K, a larger monitoring range is needed to ensure that there are K objects in this range, and more candidate objects are needed to verify. The overall time consumption of the proposed DACKNN algorithm is lower than those of other two query algorithms, and its overall performance is higher than CKNN and SKNN. Figure 20 shows the accuracy of three algorithms with the increase of *K*. It can be seen that the moving objects towards query object can always be obtained 100% by DACKNN algorithm as *K* increases, while the accuracy of moving objects obtained by SCKNN algorithm is between 60% and 80%, and the accuracy of CKNN algorithm is nearly 50%.

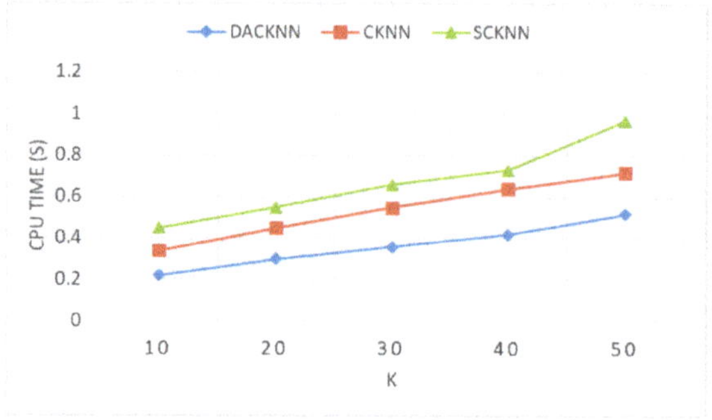

Figure 19. Time consumption of Algorithms for different K values.

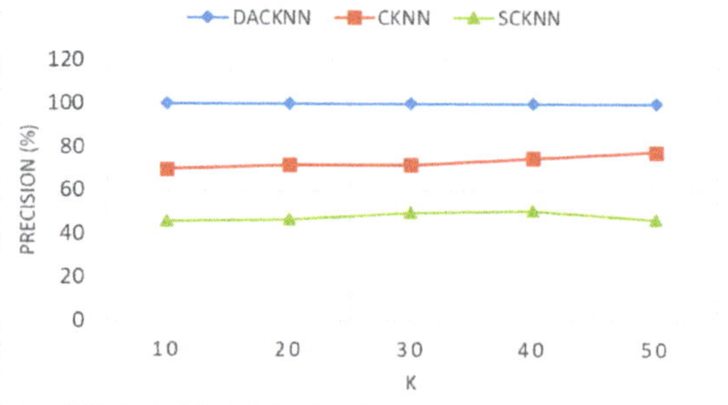

Figure 20. Accuracy of Algorithms for different K values.

The results show that the proposed DACKNN algorithm is superior to SCKNN and CKNN. By utilizing the directional characteristics of moving objects, the DACKNN determine moving objects moving towards query point, and filters out some objects far from the query point in the monitoring range, thus improving the overall performance and effectiveness of the query algorithm.

6. Conclusions

This paper presents a direction-aware continuous moving K-nearest-neighbor query algorithm in road networks. In this algorithm, we adopt an efficient direction determination method that is based on road network expansion to quickly judge whether moving objects are moving towards the query point. This method can filter out moving objects that are far away from the query point and reduce the number of candidate objects for continuous K-nearest-neighbor query, thereby improving algorithm efficiency. In this paper, we define the road network distance of a moving object from the query object as a function of time, so we can determines when candidate objects replace the objects in the result set. The algorithm guarantees the consistency of the query results within each sub-interval; hence, it does not make continuous query requests on continuous K-nearest neighbor queries, avoids repeated queries and reduces computational costs.

This paper initially solves the problem of direction-aware continuous moving K-nearest neighbor query in road networks. However, there are still some limitations in our method. On the one hand, due to the simulated moving objects information, the more moving objects there are, the more points will move outside the edge, and the information of each moving point that arrives at the specified node will be recorded in memory, which may lead to memory overflow of computers and cause substantial difficulties in the process of comparative tests. On the other hand, when the moving object reaches a road network node, an adjacent edge of the node is randomly selected as its next traveling edge;

the moving object will travel along the selected edge. This randomness does not match the reality. In future research, we will study the processing of the node and the variable motion speed of moving objects to be more realistic.

Author Contributions: Conceptualization, T.D., Y.S. and L.Y.; Methodology, Y.S. and L.Y.; Original draft preparation, Y.Y. and Y.S.; Paper writing Y.S. and L.Y.; Data quality check, Y.S. and L.Y.; Data validation, L.Z.; Paper revision, T.D. and L.Y.; Writing-revision &editing, Y.S. and L.Y.; Project Administration, T.D.

Funding: This research was supported by the following foundations: National Natural Science Foundation of China (No.61672464, No.61572437).

Conflicts of Interest: The authors declare no conflict of interest.

References

1. Tao, Y.; Papadias, D.; Shen, Q. Continuous nearest neighbor search. In Proceedings of the 28th international conference on Very Large Data Bases, Hong Kong, China, 20–23 August 2002; pp. 287–298.
2. Cho, H.J.; Chung, C.W. An Efficient and Scalable Approach to CNN Queries in a Road Network. In Proceedings of the International Conference on Very Large Data Bases, Trondheim, Norway, 30 August–2 September 2005; pp. 865–876.
3. Kolahdouzan, M.R.; Shahabi, C. Continuous K nearest neighbor queries in spatial network databases. In Proceedings of the STDBM, Toronto, ON, Canada, 30 August 2004; pp. 44–50.
4. Liu, X.; Shekhar, S.; Chawla, S. Object-based directional query processing in spatial databases. *IEEE Trans. Knowl. Data Eng.* **2003**, *15*, 295–304.
5. Patroumpas, K.; Sellis, T. Monitoring Orientation of Moving Objects around Focal Points. In Proceedings of the Symposium on Large Spatial Databases, Aalborg, Denmark, 8–10 July 2009; pp. 228–246.
6. Nutanong, S.; Tanin, E.; Zhang, R. Visible nearest neighbor queries. In Proceedings of the DASFAA, Bangkok, Thailand, 9–12 April 2007; pp. 876–883.
7. Papadias, D.; Zhang, J.; Mamoulis, N.; Tao, Y. Query processing in spatial network databases. In Proceedings of the Very Large Data Bases, Berlin, Germany, 9–12 September 2003; pp. 802–813.
8. Ahuja, R.K.; Mehlhorn, K.; Orlin, J.B.; Tarjan, R.E. Faster algorithms for the shortest path problem. *J. ACM* **1990**, *37*, 213–223. [CrossRef]
9. Kolahdouzan, M.R.; Shahabi, C. Voronoi-based k nearest neighbor search for spatial network databases. In Proceedings of the Thirtieth International Conference on Very Large Data Bases, Toronto, ON, Canada, 31 August—3 September 2004; Volume 30.
10. Hu, H.; Lee, D.L.; Xu, J. Fast nearest neighbor search on road networks. In Proceedings of the Extending Database Technology, Munich, Germany, 26–31 March 2006; pp. 186–203.
11. Hu, H.; Lee, D.L.; Lee, V.C. Distance indexing on road networks. In Proceedings of the Very Large Data Bases, Seoul, Korea, 12–15 September 2006; pp. 894–905.
12. Lee, K.C.K.; Lee, W.C.; Zheng, B. Fast object search on road networks. In Proceedings of the International Conference on Extending Database Technology, Saint Petersburg, Russia, 23–27 March 2009; pp. 1018–1029.
13. Shahabi, C.; Kolahdouzan, M.R.; Sharifzadeh, M. A road network embedding technique for K-nearest neighbor search in moving object databases. *GeoInformatica* **2003**, *7*, 255–273. [CrossRef]
14. Mouratidis, K.; Yiu, M.; Papadias, D.; Mamoulis, N. Continuous nearest neighbor monitoring in road networks. In Proceedings of the VLDB, Seoul, Korea, 12–15 September 2006; pp. 43–54.
15. Demiryurek, U.; Banaei-Kashani, F.; Shahabi, C. Efficient Continuous Nearest Neighbor Query in Spatial Networks Using Euclidean Restriction. In Proceedings of the International Symposium on Spatial and Temporal Databases, Aalborg, Denmark, 8–10 July 2009.
16. Huang, Y.K.; Chen, Z.W.; Lee, C. Continuous K-Nearest Neighbor Query over Moving Objects in Road Networks. In *Advances in Data and Web Management*; Springer: Berlin/Heidelberg, Germany, 2009; pp. 27–38.
17. Li, G.; Fan, P.; Li, Y.; Du, J. An Efficient Technique for Continuous K-Nearest Neighbor Query Processing on Moving Objects in a Road Network. In Proceedings of the Computer and Information Technology, 29 June–1 July 2010; pp. 627–634.
18. Gao, Y.; Zheng, B.; Lee, W.C.; Chen, G. Continuous visible nearest neighbor queries. In Proceedings of the EDBT, Saint Petersburg, Russia, 24–26 March 2009; pp. 144–155.

19. Li, G.; Feng, J.; Xu, J. Desks: Direction-aware spatial keyword search. In Proceedings of the IEEE 28th International Conference on Data Engineering (ICDE), Arlington, VA, USA, 1–5 April 2012; pp. 474–485.
20. Yi, S.; Ryu, H.; Son, J.; Chung, Y.D. View field nearest neighbor: A novel type of spatial queries. *Inf. Sci.* **2014**, *275*, 68–82. [CrossRef]
21. Lee, K.C.K.; Lee, W.C.; Leong, H.V. Nearest surrounder queries. In Proceedings of the ICDE, Atlanta, GA, USA, 3–8 April 2006; p. 85.
22. Lee, K.C.K.; Schiffman, J.; Zheng, B.; Lee, W.-C.; Leong, H.V. Tracking nearest surrounders in moving object environments. In Proceedings of the ICPS, Lyon, France, 26–29 June 2006; pp. 3–12.
23. Lee, K.C.K.; Schiffman, J.; Zheng, B.; Lee, W.-C.; Leong, H.V. Roundeye: A system for tracking nearest surrounders in moving object environments. *J. Syst. Softw.* **2007**, *80*, 2063–2076. [CrossRef]
24. Guo, X.; Zheng, B.; Ishikawa, Y.; Gao, Y. Direction-based surrounder queries for moving recommendations. *VLDB J.* **2001**, *20*, 743–766. [CrossRef]
25. Lee, M.J.; Choi, D.W.; Kim, S.; Park, H.M.; Choi, S.; Chung, C.W. The direction-constrained k nearest neighbor query. *GeoInformatica* **2016**, *20*, 471–502. [CrossRef]
26. Chen, Z.; Shen, H.T.; Zhou, X.; Yu, J.X. Monitoring path nearestneighbor in road networks. In Proceedings of the SIGMOD, Providence, RI, USA, 29 June–2 July 2009; pp. 591–602.
27. Tianyang, D.; Lulu, Y.; Qiang, C.; Bin, C.; Jing, F. Direction-aware KNN queries for moving objects in a road network. *World Wide Web* **2019**, *22*, 1765–1797. [CrossRef]
28. U.S. Census Bureau. TIGER/Line Shapefiles. 2013. Available online: http://www.census.gov/geo/www/tiger (accessed on 1 September 2013).
29. Brinkhoff, T. A framework for generating network-based moving objects. *GeoInformatica* **2002**, *6*, 153–180. [CrossRef]

 International Journal of *Geo-Information*

Article

The Distribution Pattern of the Railway Network in China at the County Level

Minmin Li [1,2], Renzhong Guo [1,2], You Li [1], Biao He [1,*] and Yong Fan [1]

[1]   Guangdong Key Laboratory of Urban Informatics & Shenzhen Key Laboratory of Spatial Smart Sensing and Services & Research Institute for Smart Cities, School of Architecture and Urban Planning, Shenzhen University, Nanhai Ave 3688, Shenzhen 518060, China
[2]   Polytechnic Center for Territory Spatial Big-Data, MNR of China, Lotus Pond West Road 28, Beijing 100036, China
*   Correspondence: hebiao@szu.edu.cn; Tel.: +86-0755-2697-9741

Received: 13 May 2019; Accepted: 28 July 2019; Published: 30 July 2019

Abstract: Evaluation of the railway network distribution and its impacts on social and economic development has great significance for building an efficient and comprehensive railway system. To address the lack of evaluation indicators to assess the railway network distribution pattern at the macro scale, this study selects eight indicators—railway network density, railway network proximity, the shortest travel time, train frequency, population, Gross Domestic Product (GDP), the gross industrial value above designated size, and fixed asset investment—as the basis of an integrated railway network distribution index which is used to characterize China's railway network distribution using geographical information system (GIS) technology. The research shows that, in 2015, the railway network distribution was low in almost half of China's counties and that there were obvious differences in distribution between counties in the east and west. In addition, multiple dense areas of railway network distribution were identified. The results suggest that it might be advisable to strengthen the connections between large and small cities in the eastern region and that the major urban agglomerations in the midwest could focus on strengthening the construction of railway facilities to increase the urban vitality of the western region. This study can be used to guide the optimization of railway network structures and provide a macro decision-making reference for the planning and evaluation of major railway projects in China.

Keywords: railway network; railway network density; proximity; the shortest travel time; railway network distribution index

1. Introduction

Railways are a basic element of transportation systems and have played a key role in social and economic development in many countries since the 19th century. With the recent rapid development of urbanization in China, by the end of 2016, the length of the railway network had reached 1.24×10^5 km, and the length of China's high-speed rail (HSR) network was nearly 66 times that of the United States (United States: 362 km) [1]. China's first HSR route (Shenyang–Qinhuangdao), on which trains operate at a speed of 200 km/h, opened in 2003, almost 50 years after the world's first HSR route in Japan in 1964. By July 2013, however, China had the largest HSR network in the world (at 9760 km), accounting for 46% of the world's total [2]. At the same time, China's GDP (USD 11.20 trillion) is only three-fifths that of the United States (USD 18.62 trillion) [1], which implies a potential mismatch between railway investment and economic development and highlights the significance of having an efficient and comprehensive railway system. Evaluating the railway network distribution and its impact on social and economic development is an important area of research in economic geography and regional economics [3,4]. Accessibility has long been a central issue in transport geography and is a commonly used indicator in the field of transportation network analysis, transport planning, and land use [5,6].

Accessibility is a popular measure for assessing the overall spatial structure of a transportation network [7]. In 1959, Hansen was the first to define accessibility as the size of the interactions between nodes in a transportation network, and he suggested a method to measure it in metropolitan areas [8]. In 1979, Morris stated that accessibility is the means to reach a given activity site from a certain place by a specific transportation system [9]. Since the formation of these definitions, accessibility has been a central theme in transportation studies, and its measurement can be divided into three groups based on function: spatial separation, cumulative opportunity, and spatial interaction measures [10]. The first group involves calculating the topological length, the shortest distance, time, or cost between two nodes [11,12], and only measures the connectivity of the transportation network. The second group focuses on the proximity of cities to development opportunities and involves estimating the size of the population or the scale of the economic activities that can be reached from a node within a certain period of time [13,14]. The third group comprises what are called potential values [15,16].

The advantages and disadvantages of the three groups are as follows. The methods of the first group take the cost of the individual flows in the transportation network into account, it do not consider the distance attenuation and the magnitude of the force at each point. The methods of the second group essentially measure accessibility by evaluating the convenience of a certain point of travel, without considering the interaction between the measurement point and the attraction point and the attenuation of its spatial effect with distance. The methods of the third group combine the spatial effect, distance, and gravitational scale of each attraction point in space to measure the accessibility—the larger the force between an attraction point and the measurement point, the smaller the distance between them and the higher the accessibility level.

In summary, scholars have developed research methods and applications for railway accessibility, laying a methodological foundation for the study of railway network distribution. The above methods are used to analyze the spatial effects of transportation accessibility or transportation conditions from different perspectives. However, although many scholars have macroscopically analyzed railway distribution, a method that systematically and comprehensively evaluates the railway network distribution combined with the social economy is still lacking. This is need to further refine the system mechanism of the railway network and shape the regional spatial structure.

Therefore, based on the passenger railway networks, this study constructs a railway network distribution index including assessment of the railway network density, railway network proximity, the shortest travel time, train frequency, and social-economic indicators, to explore the characteristics of China's railway network distribution in 2015. This method could be used to optimize railway network structure and as a macro-level decision-making reference for evaluating major railway projects.

2. Materials and Methods

2.1. Research Area and Data Sources

In China, the administrative divisions are divided into provincial administrative districts, prefecture-level administrative districts, county-level administrative districts, and the town-level administrative districts. The county-level administrative districts include city-governed districts, county-level cities, counties, autonomous counties, flags, autonomous flags, special zones, and forest areas [17]. In this study, county-level administrative districts are used as the basic statistical unit, and city-governed districts are merged into one.

There are many distribution structures of urban agglomerations according to different criteria and principles. In China, the main urban agglomerations structures are Shimou Yao's "6 + 7" plan [18], Chaolin Gu's "3 + 3 + 7 + 17" plan [19], and Chuanglin Fang's "5 + 9 + 6" plan [20]. In this paper, Chuanglin Fang's "5 + 9 + 6" plan is used to discuss the railway network distribution with regard to urban agglomeration, because this plan takes China's major function-oriented zoning [21] and national urban system planning for 2006–2020 into account [22] and can describe urban agglomeration comprehensively. Focused on the new urbanization policy, this plan scientifically cultivates large, medium and small scale and gradient urban agglomerations, building five national-level urban agglomerations (Yangtze River Delta, Pearl River Delta, Beijing-Tianjin-Hebei, Triangle of Central China, and Chengyu), nine regional-level urban agglomerations (Harbin-Changchun, Shandong Peninsula, Liaozhongnan, West side of the Straits, Central Plains, Guanzhong Plain, Jianghuai, Beibu Gulf, and Tianshan North Slop), and six local-level urban agglomerations (Hubaoeyu, Jinzhong, Ningxia along the Yellow River, Lanxi, Dianzhong, and Qianzhong), promotes the coordinated development of urban agglomerations at different levels, and forms a new system of "5 + 9 + 6" spatial structure of China's urban agglomerations. This plan provides a scientific prescription for rational construction of urban agglomerations, plays the high-end think tank of the development of the national urban agglomerations, promotes the urban agglomerations as the main body for the new urbanization, and also makes an important decision-making support for the construction of the urban agglomerations.

Due to China's accelerated economic and social development, the country is divided into four major economic regions: the eastern region, which includes Beijing City, Tianjin City, Hebei Province, Shanghai City, Jiangsu Province, Zhejiang Province, Fujian Province, Shandong Province, Guangdong Province, and Hainan Province; the northeast region, which includes Liaoning Province, Jilin Province, and Heilongjiang Province; the central region, which includes Shanxi Province, Anhui Province, Jiangxi Province, Henan Province, Hubei Province, and Hunan Province; and the western region, which includes Inner Mongolia, Guangxi, Chongqing City, Sichuan Province, Guizhou Province, Yunnan Province, Tibet, Shanxi Province, Gansu Province, Qinghai Province, Ningxia, and Xinjiang [23]. The main features of economic and social development in the various regions are the development of the western region, the revitalization of the northeast region, the rise of the central region, and the leading development in the eastern region.

The specific data sources, shown in Table 1, included (1) land area, population, GDP, gross industrial value above designated size, and fixed asset investment by county-level administrative districts of China for 2015, obtained from the China County Statistical Yearbook and the China City Statistical Yearbook [24,25]; and (2) 2015 vector data of the railway line network, railway stations, and train frequency, acquired from the website of https://www.amap.com/ and the Railway Customer Service Center of China using web crawler technology. Specifically, land area, population, GDP, gross industrial value above designated size, fixed asset investment, and train frequency data are spatialized and divided into each county-level administrative districts of China; the railway line network and railway station data have gone through error analysis and correction, format conversion, projection conversion, scale consistency and other processing, being prepared for the subsequent county-level data overlay analysis.

Table 1. Data sources for the railway network distribution of China in 2015.

Name	Source	Time	Method
Land area, population, GDP, gross industrial value above designated size, fixed asset investment	*China County Statistical Yearbook, China City Statistical Yearbook*	2015	Spatial visualization
Railway line network, Railway stations, train frequency	AMAP, Railway Customer Service Center of China	2015	Web crawler technology

2.2. Methods

Based on the above data, the aim of this study was to examine the distribution pattern of the railway network at the county level using a new railway network distribution index. In this study, the distribution pattern of the railway network was assessed by the railway network density; railway network proximity; the shortest travel time; train frequency, which reflect the railway frequency of each county; the population; GDP; the gross industrial value above designated size; and fixed asset investments, which provide an indication of the social economic index. These indicators comprehensively reflect the characteristics of railway network distribution from the aspects of regional support ability, external convenience degree, external accessibility, service ability, and coordination with the population and economy [20,26,27].

Based on these indicators, the railway network distribution index reflects the characteristics of railway network distribution. We have selected the analytic hierarchy process (AHP) method to assign the weight of each indicator. The main steps of the method procedure are: (1) the qualitative method is used firstly to select the indicators related to the railway network distribution, then the correlations between these indicators are analyzed, and finally eight indicators with less correlations are selected; (2) we construct the judgment matrix by Delphi expert investigation method, then solve for the weight of each indicator and judge the rationality of the weight vector. AHP provides a new, concise and practical decision-making method for the study of complex system which is composed of interrelated and mutually constrained factors [28].

This study mainly referred to the method proposed by Jin [4] with adaptations to local conditions. In his study, the concept of transportation superiority is presented from three aspects—quality, quantity, and field—to reflect the scale, technical level, and relative advantage of transport infrastructure in China. Then, the transport network density, degree of influence of the transport trunk line, and transport superiority degree of locations are used as basic indicators to express transport superiority by utilizing Geographical Information System (GIS) technology. Based on the concept of transportation superiority, we used eight indicators to indicate the railway network density, railway network proximity, the shortest travel time, train frequency, and also social-economic indicators, which are better able to describe the characteristics of railway network distribution.

The specific technical process is shown in Figure 1. The main research steps are as follows: (1) based on railway vector data, the railway network density was used to evaluate the supporting capacity of railway network facilities, and it is mainly applicable to linear transportation facilities [29,30]; (2) based on the location of the county center and railway station, the railway network proximity was used to reflect the convenience of the county with respect to other counties, the higher the proximity, the better the traffic conditions, the higher the support for regional development, and the greater the potential for external connection [31,32]; (3) based on the train frequency data, the shortest travel time was used to reflect the accessibility of the railway network, and the train frequency data also reflected the external service capacity of each railway station. In China, the railway is the main mode of transportation for medium—and long-distance passenger transportation, and the shortest travel time directly reflects the importance and connectivity of the county in the national railway network [33,34]; (4) the above indicators combine social and economic factors, such as population, GDP, gross industrial value above designated size, and fixed asset investment, to depict the pattern of the overall railway network distribution.

The specific methods for determining the distribution pattern of the railway network in China at the county level were as follows:

(1) Railway network density, C_{1i}

The railway network density can be used to evaluate the supporting ability of railway infrastructure to regional development. Railway network density is a positive indicator: the larger its value, the denser the railway network and the better the regional railway conditions. Let the railway network density of county i be C_{1i}, the length of the railway network of county i be L_i, and the area of county i be A_i; then the county railway network density can be calculated as follows:

$$C_{1i} = \frac{L_i}{A_i} \quad i = (1,2,3,\ldots,n).$$ (1)

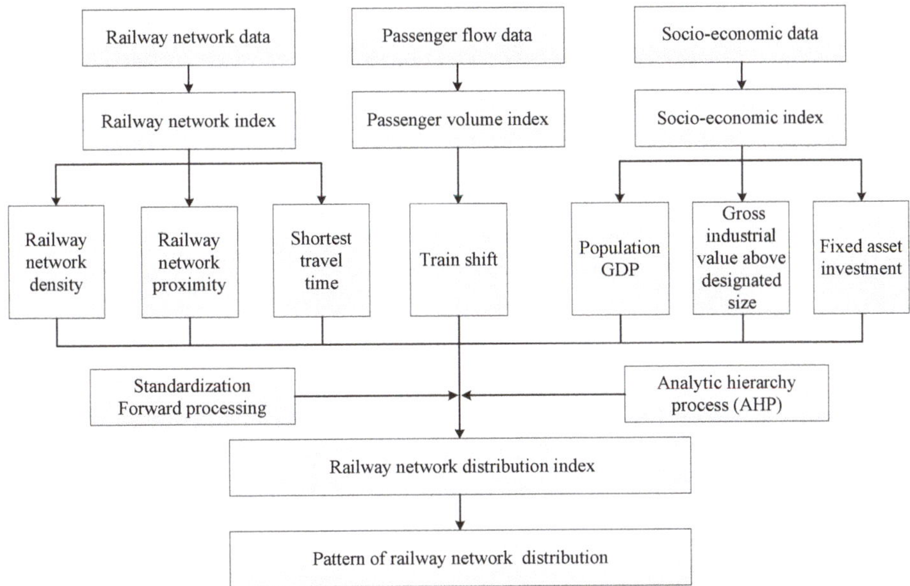

Figure 1. Technical flow for describing the distribution pattern of the railway network.

(2)　Railway network proximity, C_{2i}

This reflects the ease of traveling from a county to other counties via the proximity of its railway network. In general, proximity is used to describe the reachability of two features in geographical space. In this study, proximity was used to characterize the impact of railways on regional transportation advantages. Based on the distance from the nearest railway station to the county center, qualitative indicators were quantified using expert scoring, and then a proximity value was assigned and classified. Table 2 presents the proximity of the railway network infrastructure in county *i*, and the distance from the nearest railway station to the center of county *i* is used to determine the weight; *i* represents a county in China. The expression of the proximity is as follows:

$$C_{2i} = \sum P_i \quad i = (1, 2, 3, \dots, n).$$
(2)

Table 2. Proximity of a railway network.

Type	Standard	P_i
Railway	Own railway station	1.5
	L ≤ 30 km	1.0
	30 km < L ≤ 60 km	0.5
	L > 60 km	0.0

L is the distance from the nearest railway station to the county center. The weighted value refers to the provincial-level functional area division [35].

(3)　The shortest travel time, C_{3i}

The shortest travel time reflects the accessibility of the railway network in a region and is an important indicator to measure the external connections of a railway network. The shortest travel time is an hour, and the higher this value is, the worse the accessibility is.

This study used the train frequency data to calculate the overall shortest time in each county. T_{ij} is the shortest railway time in county i and county j, and when there was no railway connection between the two counties, it was assumed that the shortest travel time was the maximum time for the railway service that can connect county i and county j. N is the number of counties in the region. The expression of the shortest travel time is as follows:

$$C_{3i} = \sum_{j=1}^{n} T_{ij}/n \quad i = (1, 2, 3, \ldots, n).$$ (3)

(4) Train frequency, C_{4i}

Train frequency data can reflect the railway's service to the county area. i represents the county of China, while m_i is the number of train frequency through county i:

$$C_{4i} = m_i \quad i = (1, 2, 3, \ldots, n).$$ (4)

(5) Railway network distribution index, C

We integrated the railway network density C_{1i}, railway network proximity C_{2i}, the shortest travel time C_{3i}, the train frequency C_{4i}, socioeconomic indicators of the population C_{5i}, the GDP C_{6i}, the gross industrial value above designated size C_{7i}, and fixed asset investment C_{8i}, through the analytic hierarchy process (AHP) method and, combined with the existing research results [4], a railway network distribution index model was constructed. First, as an intermediate step, the eight indicators were standardized; as shown in Equation (5). C_{ji} is the normalized value of index j in county i, E_{ji} is the original value of index j in county i, $Max(E_{ji})$ is the maximum value of index j in county i, and $Min(E_{ji})$ is the minimum value of index j in county i ($j = 1, 2, 3, 4, 5, 6, 7, 8, i \in (1, 2, 3, \ldots, n)$).

$$C_{ji} = \frac{E_{ji} - Min(E_{ji})}{Max(E_{ji}) - Min(E_{ji})}.$$ (5)

Secondly, the inverse indicator, such as the shortest travel time of the railway, was used to forward the reverse index. Thirdly, according to the degree of action of each indicator in the railway network distribution, the judgment matrix was constructed, and the weight of each indicator a_j was calculated by AHP.

Hence, the railway network distribution index, C, of each county i can then be expressed as

$$C = \sum_{j=1}^{8} a_j \times C_{ji} (j = 1, 2, 3)$$ (6)

where a_j is the weight of each indicator, determined using the AHP.

3. Results

3.1. Spatial Pattern of Railway Network Density

China's railways, which are divided into ordinary and high-speed railways, currently provide important support for regional external relations. Figure 2 shows the distribution pattern of the railway network density based on the county-level administrative regions of China and reflect the ability of railway facilities to support the development of a county.

The following features of the distribution pattern of the railway network density are noteworthy: (1) The total length of railways in China is 77,800 km, and the total railway network density is 0.0082 km/km^2. According to medium- and long-term railway network planning (2016-2030), by 2025, the railway network needs to reach about 175,000 km, of which the high-speed railway will account for about 38,000 km. (2) Railway connections are not found in 44.79% of counties. The lengths of railways are relatively short in some counties, so their densities are also low. For example, the railway network density between 0 and 0.01 km/km^2 in 195 counties (8.83% of all counties), and such counties have low support capacity for development. The railway network density is between 0.01 and 0.03 km/km^2 in 617 counties (27.94% of total counties); between 0.03 and 0.05 km/km^2 in 253 counties (11.46%). Only 6.97% of counties have railway network densities higher than 0.05 km/km^2, and the railway contributes significantly to the development of these counties. (3) Railway network density is high in the central and eastern parts of China, while it is relatively low in the western part of China. In eastern China, in particular, the railway network density of the south is lower than that of the north. (4) Some regions of the railway network are particularly dense, including the Beijing-Tianjin region, the Yangtze River Delta region, and the Zhengzhou, Wuhan, and Chengdu Economic Zone. These areas have high spatial coupling with economic agglomeration and an urban system [36]; thus, railway networks have a significant ability to support the socioeconomic development of China's dense urban areas.

Generally speaking, the development of China's railway network has supported the rapid urbanization process to a certain extent, it based on the considerable population and economic scale, the construction of the railway network needs to be strengthened further to realize national railway connectivity, high-speed railways between provincial capitals, the rapid arrival of intercity railways, and the distribution of basic coverage to all counties by 2030 [37].

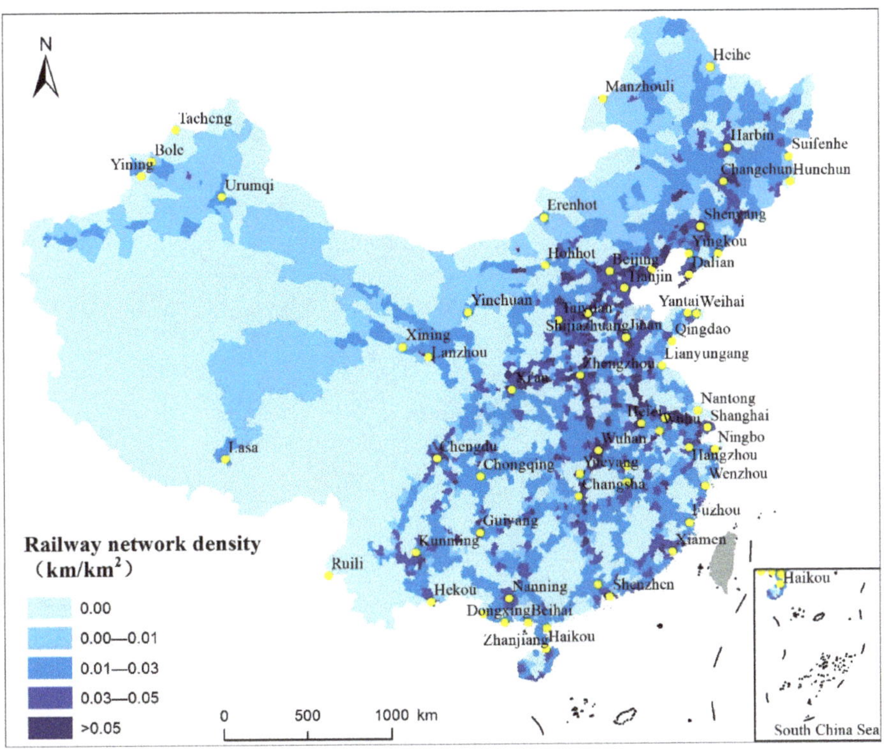

Figure 2. The spatial pattern of China's railway network density in 2015.

3.2. Spatial Pattern of Railway Network Proximity

The railway network proximity can reflect the external accessibility and convenience of travel of a county. The spatial pattern of railway network proximity at the county level is shown in Figure 3, and has the following main characteristics: (1) The railway network proximity is 0 in 369 counties (16.71% of all counties), indicating a lack of accessibility. The number of counties with a railway network proximity of 0–0.5, and thus a lower supporting capacity for development, is 486 (22.01%

of the total). The railway network proximity of 168 counties is in the range 0.5–1.0 (7.61% of all counties), suggesting a positive effect of the railway network on the development of these counties. The highest railway network proximity is found in 53.67% of the counties in China, which indicates that the railway network plays an important role in the development of these counties. (2) The railway network proximity is high in the central and eastern parts of China, it relatively low in western China, which indicates that railway network proximity has a stronger supporting capacity in eastern counties than in western counties. (3) In eastern China, there is a clear spatial difference in railway network proximity between the north and the south. For example, the northeastern region, Inner Mongolia, the capitals of Ningxia and Qinghai provinces, Beijing-Tianjin-Hebei, the Yangtze River Delta, Wuhan, and Chongqing have a high railway network proximity, while the southern region has a low railway network proximity and presents a strip distribution.

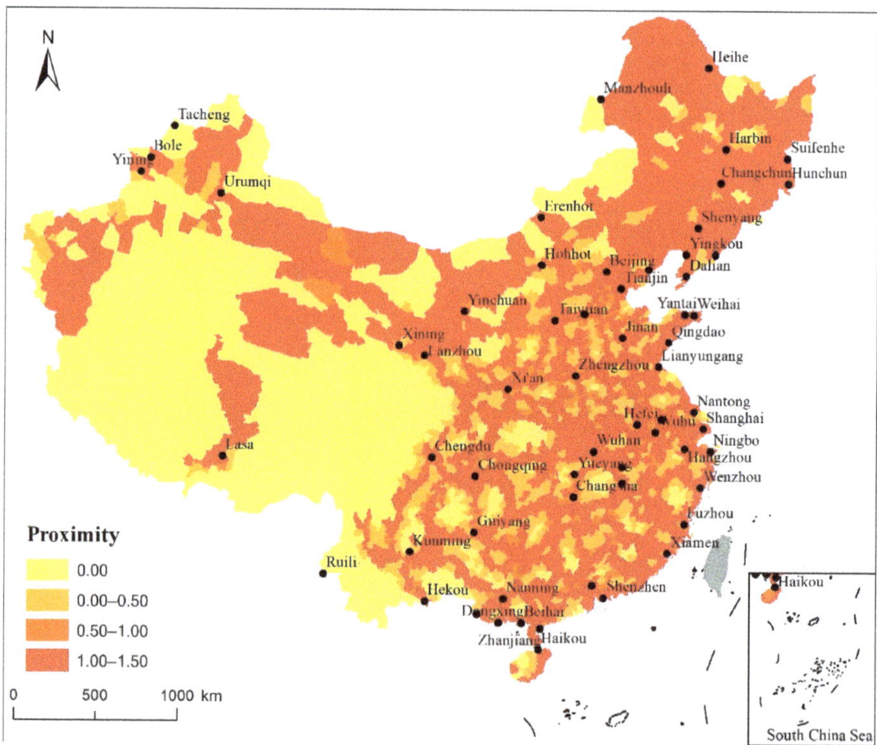

Figure 3. The spatial pattern of China's railway network proximity in 2015.

3.3. Spatial Pattern of the Shortest Travel Time

The shortest travel time shows the external connectivity of the railway network and is an important indicator to measure the regional railway network structure and external geographical connections. The spatial pattern of the shortest travel time at the county level is shown in Figure 4, and has the following main characteristics: (1) The shortest travel time is less than 9 h in 304 counties (13.77%), indicating that these counties have the highest degree of ease of external connection. The shortest travel time is 9–12 h in 594 counties (26.90%; higher degree of ease of external connection); 12–15 h in 166 counties (7.52%; lower degree of ease of external connection); and greater than 15 h in 1144 counties (51.81%; lowest degree of ease of external connection). (2) The shortest travel time has significant spatial differences throughout the country, characterized by a gradual rise from the coast to inland, and a concentration of areas with the shortest travel time in the eastern region.

(3) Hachang agglomeration, Beijing-Tianjin-Hebei, the Yangtze River Delta (Shanghai, Nanjing, Hefei, and Hangzhou), the Pearl River Delta, Wuhan, Nanchang, and Changsha have the shortest travel time, and cover a wide area.

Generally speaking, based on the shortest travel time, the convenience of the railway network in China is generally low. On the one hand, this is due to restrictions placed on the construction of railways by the natural terrain; on the other hand, because the national railway system is not perfect, it is still necessary to strengthen the railway links and exchanges within the inter-regional and urban agglomerations.

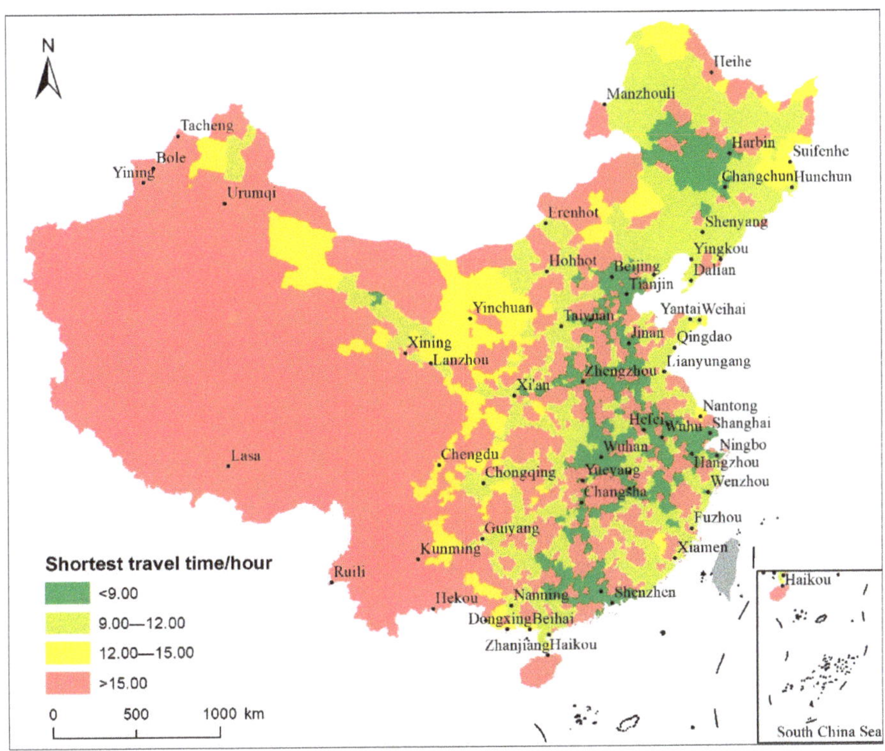

Figure 4. Spatial pattern of the shortest travel time by railway of China in 2015.

3.4. Spatial Pattern of Railway Network Distribution

The railway network density, railway network proximity, and the shortest travel time reflect the contribution of the railway network to each county's development, and demonstrate the counties' future development potential. Integrating these three indicators and combining train frequency, population, GDP, gross industrial value above designated size, and fixed asset investment, the spatial pattern of the railway network in China was determined by setting the weight of each factor using the AHP method, as shown in Figure 5. The weights of each indicator obtained by the AHP method were 0.1569, 0.2273, 0.3268, 0.1077, 0.0242, 0.0734, 0.0498, and 0.0340, respectively, and the consistency ratio was found to be CR = 0.03 < 0.1 which was through the consistency test. Hence, the normalized eigenvectors can be used as weight vectors.

The railway network distribution index is divided into five levels, namely, 0–0.01, lacking railway network distribution; 0.01–0.20, relatively lacking in railway network distribution; 0.20–0.35, moderate railway network distribution; 0.35–0.45, good railway network distribution; and >0.45, perfect railway network distribution. The following are the main observations of the railway network distribution in China: (1) 660 counties (29.89%) have a perfect railway network distribution, which provides the strongest support for future development; 423 counties (19.16%) have a good railway network distribution, indicating strong support for future development; 99 counties (4.48%) have moderate railway network distribution, and railway facilities in these areas have general support capacity but have strong development potential and opportunities; and 1026 counties (46.47%) lack railway network distribution and do not provide enough support for future development. (2) The railway network distribution index shows a pattern of decline from coastal counties toward the interior of the country, with the highest railway network distribution indices in coastal counties and the lowest indices in western counties and parts of central counties. (3) The railway network distribution index is significantly higher in central and eastern counties than in western counties, for example, Beijing-Tianjin-Hebei, the Yangtze River Delta, the Pearl River Delta, Chengdu-Chongqing, the Wuhan metropolitan area, the North Gulf Area, Shandong Peninsula, and Hachang District have higher railway network distributions than other areas of China, and also have strong development support.

Generally speaking, the railway network distribution of China is relatively effective, especially in the eastern region, where a relatively complete railway connection network exists, while the western region has a relatively inadequate network distribution. In the future, it might be advisable to strengthen the connections between large cities and small cities in the eastern region, so that large cities can drive the development of small cities around them. The major urban agglomerations in the central and western areas also could focus on strengthening the construction of railway facilities to promote the movement of people and industries between the east and west regions, thereby increasing the urban vitality of the western region.

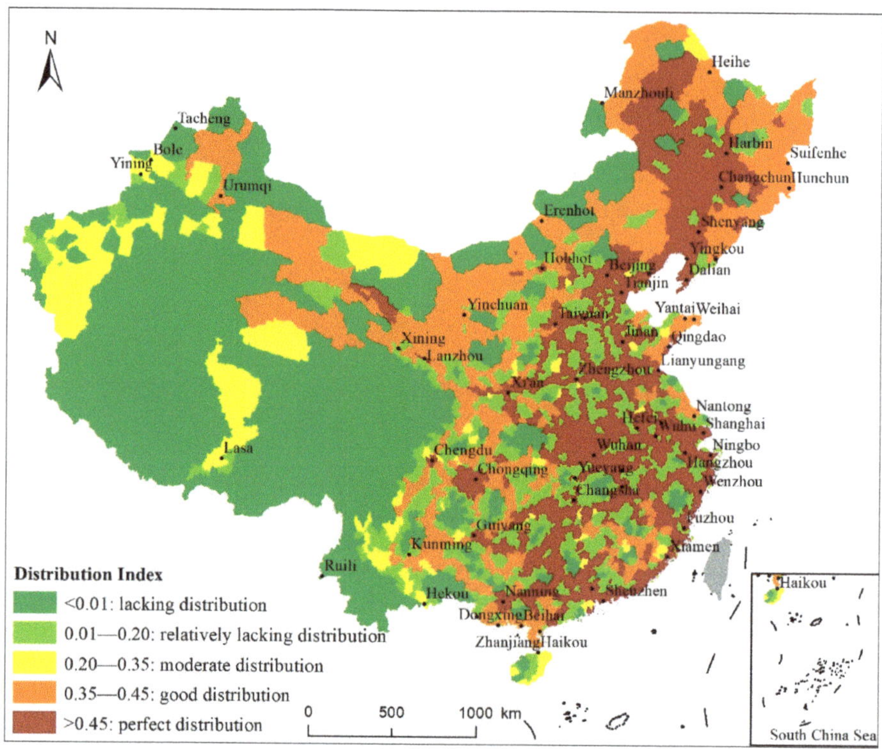

Figure 5. Spatial pattern of the railway network of China in 2015.

The distribution of railway network and economic development are mutually influential. The construction of railway infrastructure would promote the economic development of the region. Similarly, the economic development would also affect the operation and development of the railway infrastructure of the region. In this study, it is reasonable to take population and economic factors as indicators that reflect the characteristics of railway network distribution. Many scholars have studied the relationship between transportation infrastructure network and economic development, which was the 'chicken and egg' problem. For example, Xie and Levinson (2009) provided an overview of transportation networks following five main streams: network growth in transport geography; traffic flow, transportation planning, and network growth; statistical analyses of network growth; economics of network growth; and network science [38]. The review pointed out the positive interaction between economic development and transportation network construction. Levinson (2007) examined the changes that occurred in the rail network and density of population in London during the 19th and 20th centuries, and the research found that there was a positive feedback effect between population density and network density [39].

4. Discussion

4.1. Railway Network Distribution in Urban Agglomerations

The distribution structure of urban agglomerations according to the "5 + 9 + 6" plan is shown in Figure 6, in which pink areas represent five national-level urban agglomerations, yellow areas represent nine regional-level urban agglomerations, and green areas represent six local-level urban agglomerations.

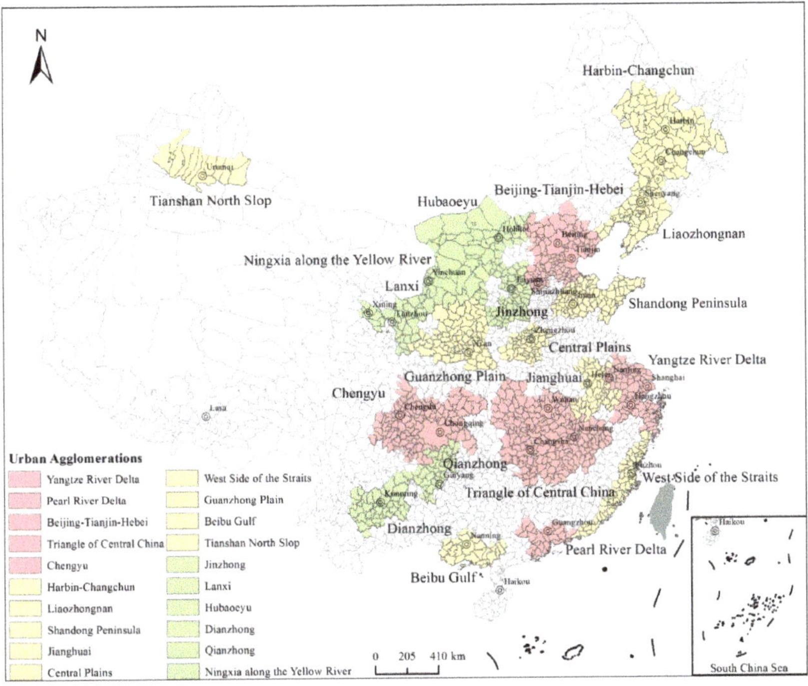

Figure 6. Distribution structure of urban agglomeration following the "5 + 9 + 6" plan of China in 2015.

Based on this distribution structure, the railway network distribution index in each urban agglomeration can be calculated (see Figure 7). The railway network distribution indices range from 0.16 to 0.41 for all urban agglomerations, indicating relatively lacking, moderate, and good distributions according to the spatial patterns of railway network distribution discussed in Section 3.4. The five national-level urban agglomerations all have moderate railway network distributions, which indicates that a moderate network distribution should continue to be a focus for supporting the fastest urbanization in national-level urban agglomerations [40,41]. Of the nine regional-level urban agglomerations, Liaozhongnan, Jianghuai, Harbin-Changchun, and the Central Plains have good railway network distributions; the west side of the Straits, Shandong Peninsula, Guanzhong Plain, and Beibu Gulf have moderate railway network distributions; and Tianshan North Slope has a relatively insufficient railway network distribution. The construction of railway network facilities should be increased as an important driving force for urbanization in these areas [42]. Of the six local-level urban agglomerations, only Qianzhong has a relatively insufficient railway network distribution, and the others have a moderate railway network distribution. These local-level urban agglomerations are mainly located in western China, with relatively low urbanization levels and a moderate railway infrastructure that could support the economic development in these areas [43].

Overall, the railway network distribution in internal urban agglomerations is not perfect; only four regional-level urban agglomerations have good railway network distributions, while the rest have medium or relatively insufficient railway network distributions. Therefore, in order to meet China's urbanization needs, as well as match the development of the population and economy, it is better to focus on strengthening the construction of railway network within the national-level urban agglomerations.

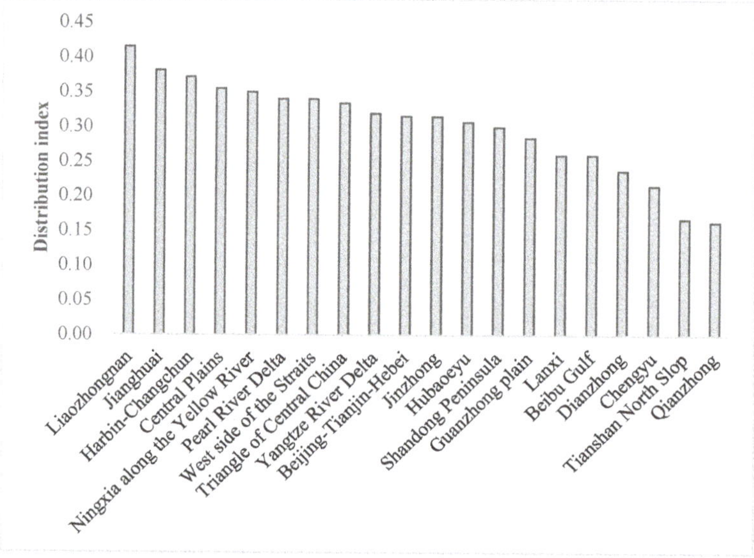

Figure 7. Railway network distribution index in each urban agglomeration in 2015.

4.2. Railway Network Distribution in Four Economic Regions

The four major economic regions of China are shown in Figure 8. The main features of economic and social development in the various regions are the development of the western region, the revitalization of the northeast region, the rise of the central region, and the leading development in the eastern region.

The railway network distribution index is 0.31 in the eastern region, 0.32 in the central region, 0.19 in the western region and 0.39 in the northeast region. The railway network distribution of the northeast region is the best, followed by those of the central and eastern regions, while the western region is the most insufficient. Wang et al. (2009) divided China's railway network expansion over the last century into four eras: preliminary construction, network skeleton, corridor building, and deep intensification [7]. Prior to 1950, the railway network mileage of the northeast region accounted for more than 40% of the network in China. In the 1970s, China implemented the policy of "reform and opening up" and shifted the economic center and railway construction to the southeast coastal region. During this period, the railway network in the eastern coastal region developed rapidly, which led to unbalanced inter-regional development. In the late 1990s, China implemented balanced regional development and successfully put forward the strategies of "the great western development strategy," "revitalizing the old industrial bases in northeast China," and "the rise of the central region", which gradually enhanced the construction of the railway network in the central and western regions. In 2015, China proposed "The Belt and Road" development strategy, which provides targeted guidance for the future distribution of the railway network.

Therefore, according to the national development strategy, the future distribution of the railway network could focus on the central and eastern regions as well as the western urban agglomerations, while the western and northeast border areas may mainly consider their national security significance. A logical distribution of the railway network will guide urbanization and the development of population and economic aggregation.

Figure 8. The four major economic regions of China.

4.3. Limitations and Future Improvements

This study mainly considers railway passenger transport, and represents the size of railway passenger transport with train frequency data. Due to the difficulty in data acquiring, we do not take into account the freight volume situation of China in this study. Adding the freight transport information will definitely make the railway network distribution pattern more accurate.

Lim and Thill (2008) investigated how the intermodal freight-transportation network affects the ability of regions to position themselves more effectively in the national space economy, and the performance of the intermodal freight network was evaluated by comparing accessibility measures based on the highway network and on the intermodal network, respectively [44]. This research can provide reference and guidance for the future research for combining passenger and freight status, and reflecting the characteristics of the railway network distribution of China more precisely.

In this study, the distance from the nearest railway station to the county center, based on the provincial-level functional area division, is used to determine the weight of railway network proximity using expert scoring. Potential future improvements could include setting different distance intervals and selecting different evaluation methods for expert scoring. In addition, the current research has only been applied to examine railway network distribution at the macro scale. The theory and evaluation method needs further testing if being applied to regions at small—and medium-scales.

5. Conclusions

Based on indicators of the railway network density, railway network proximity, the shortest travel time, train frequency, and also social-economic indicators, this study took advantage of a railway network distribution index to investigate the railway network distribution pattern in China in 2015. The results of this study could guide the optimization of the railway network structure and provide a basis for macro decision-making for the planning and evaluation of major railway infrastructure construction.

The findings were as follows: (1) In 2015, railway network density was relatively low in almost half of the counties of China. The railway network density was high in the central and eastern parts of China and relatively low in western China, and there were multiple dense railway network areas. The railway network proximity was high in nearly half of the counties in China, as was evident in the strip distribution. Based on the shortest travel time, the convenience of most counties in terms of external connections to the rest of China is generally low, with the convenience gradually rising from coastal to inland areas; the areas with the shortest travel time were mainly concentrated in the eastern region. (2) In 2015, the railway network distribution in nearly one-third of the counties of China was ideal. The distribution was grouped into two major zones with multiple dense areas. It might be advisable to strengthen the connections between large and small cities in the eastern region, so that large cities can drive the development of small cities around them. The major urban agglomerations in the midwest also could focus on several aspects such as strengthening the construction of railway facilities, promoting the movement of people and industries between the east and west regions, and increasing the urban vitality of the western region.

This study revealed the distribution pattern of the railway network in China and has significance for the optimization of the railway network structure and the development of urbanization in China. Based on the railway network distribution indicators, the national railway network is divided into five levels: insufficient railway network distribution, relatively insufficient railway network distribution, moderate railway network distribution, good railway network distribution, and perfect railway network distribution. Utilizing the advantages of the railway should be considered when planning economic development and industry activities. The five national-level urban agglomerations—Beijing-Tianjin-Hebei, the Yangtze River Delta, the Pearl River Delta, Triangle of Central China, and Chengdu-Chongqing—have moderate railway infrastructures, and local governments may harness the advantages of other transportation facilities in these regions, such as roads, aviation, and ports, to compensate for the inadequacy of the railway network, and actively participate in international economic competition to enhance China's urbanization. There are four regional-level agglomerations that have good railway network distribution—Liaozhongnan, Jianghuai, Harbin-Changchun, and Central Plains—Where the government may take full advantage of the railway network and promote regional integration. The railway network distribution of other regional-level urban agglomerations and local-level urban agglomerations is moderate or relatively lacking. In the future, the local government could plan for the reasonable construction of railway facilities and combine them with other modes of transportation to match the development of the population and economy.

Author Contributions: You Li conceptualized this research and contributed extensively to data curation and formal analysis; Minmin Li contributed to the methodology, investigation, and writing of this research; Renzhong Guo and Biao He gave ideas to validate the conceptualization and methodology; Yong Fan provided many suggestions for improving and modifying this article.

Funding: This research was funded by the China Postdoctoral Science Foundation, grant number: 2018M633108, 2018M643150 and the Natural Science Foundation of China, grant number: 41701187.

Acknowledgments: The authors are grateful for the "Connection value of transportation network based on breaking points model", "Research on Pole-like Furniture Recognition in Mobile Laser Scanning Data Considering the Topological Structure Between Components" and "Research on Spatial Structure and Evolution Model of Urban System based on Entropy Theory" projects of Shenzhen University. We also would like to thank Jian Sun, who works at Jilin University and gave many suggestions for improving this article.

Conflicts of Interest: The authors declare no conflict of interest.

References

1. The World Bank. World Development Indicators. Available online: https://data.worldbank.org (accessed on 17 April 2019).
2. UIC. High Speed Line in The World. Available online: http://www.uic.org/spip.php?mot8 (accessed on 17 April 2019).
3. Rodrigue, J.P.; Comtois, C.; Slack, B. *The Geography of Transport Systems*; Routledge: New York, NY, USA, 2009.
4. Jin, F.; Wang, C.; Li, X.; Wang, J. China's regional transport dominance: Density, proximity, and accessibility. *J. Geogr. Sci.* **2010**, *20*, 95–309. [CrossRef]
5. Geurs, K.T.; Van Wee, B. Accessibility evaluation of land-use and transport strategies: Review and research directions. *Transp. Geogr.* **2004**, *12*, 127–140. [CrossRef]
6. Monzon, A.; Lopez, E.; Ortega, E. Has HSR improved territorial cohesion in Spain? An accessibility analysis of the first 25 years: 1990–2015. *Eur. Plan. Stud.* **2019**, *27*, 1–20. [CrossRef]
7. Wang, J.; Jin, F.; Mo, H.; Wang, F. Spatiotemporal evolution of China's railway network in the 20th century: An accessibility approach. *Transp. Res. Part A Policy Pract.* **2009**, *43*, 765–778. [CrossRef]
8. Hansen, W.G. How accessibility shapes land use. *J. Am. Plan. Assoc.* **1959**, *25*, 73–76. [CrossRef]
9. Morris, J.M.; Dumble, P.L.; Wigan, M.R. Accessibility indicators for transport planning. *Transp. Res. Part A General.* **1979**, *13*, 91–109. [CrossRef]
10. Jiao, J.; Wang, J.; Jin, F.; Dunford, M. Impacts on accessibility of China's present and future HSR network. *J. Transp. Geogr.* **2014**, *40*, 123–132. [CrossRef]
11. Bruinsma, F.; Rietveld, P. The accessibility of European cities: Theoretical framework and comparison of approaches. *Environ. Plan. A* **1998**, *30*, 499–521. [CrossRef]
12. Beria, P.; Debernardi, A.; Ferrara, E. Measuring the long-distance accessibility of Italian cities. *J. Transp. Geogr.* **2017**, *62*, 66–79. [CrossRef]
13. Black, J.; Conroy, M. Accessibility measures and the social evaluation of urban structure. *Environ. Plan. A* **1977**, *9*, 1013–1031. [CrossRef]
14. Yang, L.; Zhang, X.; Hong, S.; Lin, H.; Cheng, G. The impact of walking accessibility of public services on housing prices: Based on the cumulative opportunities measure. *South China J. Econ.* **2016**, *1*, 57–70.
15. Dou, K.; Zhan, Q. Accessibility analysis of urban emergency shelters: Comparing gravity model and space syntax. In Proceedings of the 2011 International Conference on Remote Sensing, Environment and Transportation Engineering, Nanjing, China, 24–26 June 2011.
16. Franke, D.; Vorel, J.; Peltan, T. Job accessibility modelling in Prague Functional Urban Area. In Proceedings of the 2017 Smart City Symposium Prague (SCSP), Prague, The Czech Republic, 25–26 May 2017.
17. Ministry of Civil Affairs of the People's Republic of China. *Administrative Division of the People's Republic of China*; Chinese Social Publishing House: Beijing, China, 2017.
18. Yao, S.M. *Urban Agglomerations of China*; University of Science and Technology of China Press: Hefei, China, 2006.
19. Gu, C.L.; Yu, T.F.; Li, W.M. *Urbanization in China: Pattern, Process and Mechanism*; Science Press: Beijing, China, 2008.
20. Fang, C.L.; Bao, C.; Ma, H.T. *Urban Agglomeration Development in China*; Science Press: Beijing, China, 2016.
21. Fan, J.; Li, P.X. The scientific foundation of major function-oriented zoning in China. *J. Geogr. Sci.* **2007**, *19*, 515–531. [CrossRef]
22. Wang, G.T. *National Urban System Planning 2006–2020*; The Commercial Press: Beijing, China, 2010.

23. National Bureau of Statistics of China. Available online: http://www.stats.gov.cn/ztjc/zthd/sjtjr/dejtjkfr/tjkp/201106/t20110613_71947.htm (accessed on 17 April 2019).
24. National Bureau of Statistics of China. *China County Statistical Yearbook*; China Statistics Press: Beijing, China, 2016.
25. National Bureau of Statistics of China. *China City Statistical Yearbook 2016*; China Statistics Press: Beijing, China, 2016.
26. Li, M.; Guo, R.; Li, Y.; He, B.; Chen, Y.; Fan, Y. Distribution Characteristics of the Transportation Network in China at the County Level. *IEEE Access* **2019**, *7*, 49251–49261. [CrossRef]
27. Hu, H.; Wang, J.; Jin, F.; Ding, N. Evolution of regional transport dominance in China 1910–2012. *J. Geogr. Sci.* **2015**, *25*, 723–738. [CrossRef]
28. Xu, S.B. *The Principle of Analytic Hierarchy Process—Practical Decision-Making Method*; Tianjin University Press: Tianjin, China, 1988.
29. Erath, A.; Löchl, M.; Axhausen, K.W. Graph-theoretical analysis of the Swiss road and railway networks over time. *Netw. Spat. Econ.* **2009**, *9*, 379–400. [CrossRef]
30. Huang, Y.; Lu, S.; Yang, X.; Zhao, Z. Exploring railway network dynamics in China from 2008 to 2017. *ISPRS Int. J. Geo-Inf.* **2018**, *7*, 320. [CrossRef]
31. Kasraian, D.; Maat, K.; Vaan Wee, B. The impact of urban proximity, transport accessibility and policy on urban growth: A longitudinal analysis over five decades. *Environ. Plan. B Urban Anal. City Sci.* **2017**, *1*, 1–18. [CrossRef]
32. Blainey, S.P.; Armstrong, J.; Smith, A.S.; Preston, J.M. New routes on old railways: Increasing rail's mode share within the constraints of the existing railway network. *Transportation* **2016**, *43*, 425–442. [CrossRef]
33. Huber, S.; Rust, C. Calculate travel time and distance with OpenStreetMap data using the Open Source Routing Machine (OSRM). *Stata J.* **2016**, *16*, 416–423. [CrossRef]
34. Zhang, D.; Chow, C.Y.; Liu, A.; Zhang, X.; Ding, Q.; Li, Q. Efficient evaluation of shortest travel-time path queries through spatial mashups. *GeoInformatica* **2018**, *22*, 3–28. [CrossRef]
35. Chinese Academy of Sciences. *Provincial-Level Functional Area Division*; Science and technology of China press: Beijing, China, 2008.
36. Li, M.; He, B.; Guo, R.; Li, Y.; Chen, Y.; Fan, Y. Study on population distribution pattern at the county level of China. *Sustainability* **2018**, *10*, 3598. [CrossRef]
37. National Development and Reform Commission. The Issuance of the Medium-and Long-Term Railway Network Plan 2016. Available online: http://www.ndrc.gov.cn/zcfb/zcfbtz/201607/t20160720_811696.html (accessed on 17 April 2019).
38. Xie, F.; Levinson, D. Modeling the growth of transportation networks: A comprehensive review. *Netw. Spat. Econ.* **2009**, *9*, 291–307. [CrossRef]
39. Levinson, D. Density and dispersion: The co-development of land use and rail in London. *J. Econ. Geogr.* **2007**, *8*, 55–77. [CrossRef]
40. Li, X.; Huang, B.; Li, R.; Zhang, Y. Exploring the impact of high speed railways on the spatial redistribution of economic activities-Yangtze River Delta urban agglomeration as a case study. *J. Transp. Geogr.* **2016**, *57*, 194–206. [CrossRef]
41. Duan, G.Y.; Gong, H.L.; Liu, H.H.; Zhang, Y.Q.; Chen, B.B.; Lei, K.C. Monitoring and analysis of land subsidence along Beijing-Tianjin Inter-city railway. *J. Indian Soc. Remote Sens.* **2016**, *44*, 915–931.
42. Li'e, S.B.W. Spatial Pattern & Quantitative Relationship of Industrial Structure of Shandong Peninsula Urban Agglomeration. In Proceedings of the 3rd International Conference on Information Science and Control Engineering (ICISCE), Beijing, China, 8–10 July 2016.
43. Zhou, J.; Yang, L.; Li, L. The implications of high-speed rail for Chinese cities: Connectivity and accessibility. *Transp. Res. Part A Policy Pract.* **2018**, *116*, 308–326.
44. Lim, H.; Thill, J.C. Intermodal freight transportation and regional accessibility in the United States. *Environ. Plan. A* **2008**, *40*, 2006–2025. [CrossRef]

International Journal of
Geo-Information

Article

Data-driven Bicycle Network Analysis Based on Traditional Counting Methods and GPS Traces from Smartphone

Federico Rupi *, Cristian Poliziani and Joerg Schweizer

DICAM- University of Bologna, Viale Risorgimento 2, 40136 Bologna, Italy
* Correspondence: federico.rupi@unibo.it; Tel.: +39-51-209-3345

Received: 4 June 2019; Accepted: 24 July 2019; Published: 25 July 2019

Abstract: This research describes numerical methods to analyze the absolute transport demand of cyclists and to quantify the road network weaknesses of a city with the aim to identify infrastructure improvements in favor of cyclists. The methods are based on a combination of bicycle counts and map-matched GPS traces. The methods are demonstrated with data from the city of Bologna, Italy: approximately 27,500 GPS traces from cyclists were recorded over a period of one month on a volunteer basis using a smartphone application. One method estimates absolute, city-wide bicycle flows by scaling map-matched bicycle flows of the entire network to manual and instrumental bicycle counts at the main bikeways of the city. As there is a fairly high correlation between the two sources of flow data, the absolute bike-flows of the entire network have been correctly estimated. Another method describes a novel, total deviation metric per link which quantifies for each network edge the total deviation generated for cyclists in terms of extra distances traveled with respect to the shortest possible route. The deviations are accepted by cyclists either to avoid unpleasant road attributes along the shortest route or to experience more favorable road attributes along the chosen route. The total deviation metric indicates to the planner which road links are contributing most to the total deviation of all cyclists. In this way, repellant and attractive road attributes for cyclists can be identified. This is why the total deviation metric is of practical help to prioritize bike infrastructure construction on individual road network links. Finally, the map-matched traces allow the calibration of a discrete choice model between two route alternatives, considering distance, share of exclusive bikeway, and share of low-priority roads.

Keywords: GPS traces; cycling volumes; cyclists' counts; cycling network; total deviation metric; route choice model

---

1. Introduction

In recent years, due to congestion, air pollution, climate change, energy scarcity, and physical inactivity, an increasing importance has been attributed to sustainable transport modes, and in particular to cycling. Municipalities have drawn attention to these issues and started to implement different strategies to encourage a greater usage of bicycles on urban streets and to reduce car trips. Many cities have decided to invest in the construction of quality bikeways with the intention to incentivize people to cycle even medium (and long) distances on a daily basis.

Several studies have found positive correlations between bike facilities and levels of bicycling. Dill and Carr [1] have analyzed data from 35 large cities across the U.S., finding that cities with higher levels of bicycle infrastructure were characterized by higher levels of bicycle commuting. Pucher and Buehler found that the key to achieving high levels of cycling in Dutch, Danish, and German cities appears to be the provision of separate cycling facilities [2].

Many researchers have studied cyclists' preferences and have estimated route choice models for planning new bicycle infrastructure. Most of these studies were based on stated preference (SP) data and on opinion surveys. Dill and Voros [3] explored the relationships between levels of cycling and demographics, objective environmental factors, perceptions of the environment, and attitudes based on the results from a random phone survey of adults in the Portland, Oregon. Stinson and Bhat [4], proposed empirical models to evaluate the importance of factors (such as travel time and pavement quality) affecting commuter bicyclists' route choices. Winters et al. [5] evaluated 73 motivators and deterrents of cycling based on a survey (telephone interviews) and a self-administered survey (either via the web or mail) of 1402 current and potential cyclists in metropolitan Vancouver. These studies highlighted the cyclists' preference for physically separated bike paths and on-road bike lanes separated by markers. The known limitation of SP studies arises from the difference between stated and observed behavior [6].

In the past, only a few studies were based on revealed preference (RP) data due to the limited availability of this data type. The results of Howard and Burns [7] show that bicycle commuters in metropolitan Phoenix respond to the provision of bicycle facilities. Nelson and Allen [8] found that each additional mile of bikeway per 100,000 people is associated with a 0.069% increase in bicycle commuting.

Data on cycling volumes is the result of choices actually made by cyclists in a real outdoor environment and therefore help to support decision making. Such information is necessary to understand what factors influence ridership. Bike flows can be collected with traditional manual or instrumental counts, which have some drawbacks: Traditional manual counts lack spatial detail and temporal coverage [9,10]. Instrumental and permanent counting stations do provide continuous data, but cover typically only a small number of road sections [11].

More recently, the widespread use of smartphones and mobile applications for self-localization and navigation has increased the availability of observed cyclists' data in the form of time series of GPS points, called GPS traces. This type of data provides detailed information about the origin/destination of trips as well as the chosen routes. Furthermore, GPS traces allow to determine the total deviation (detour) generated for cyclists in terms of extra distances traveled with respect to the shortest possible route. Empirical data on detour rates do exist. Detours have been calculated in different ways, see for example Pritchard et al. [12] and Griffin and Jiao [13].

The availability of GPS traces led to the development of new cyclists' route choice models. Dill collected data on bicycling behavior from 166 regular cyclists (1955 trips) in the Portland, Oregon, using GPS devices [14]. This study highlighted that a well-connected network of low traffic streets may be more effective than adding bike lanes on major streets with a high volumes of motor vehicle traffic. Menghini et al. estimated the route choice model for bicyclists from a large sample of GPS observations (2498 trips) collected in Zürich, Switzerland [15]. Their conclusion has been that the trip-length dominates the choices of the Zurich cyclists. Hood et al. analyzed GPS traces from cyclists (366 users and 2777 trips) in San Francisco, USA, and proved a preference for separated bicycle lanes—especially for infrequent cyclists [16]. Broach et al. estimated the route choice of cyclists (164 users and 1449 trips) in Portland metropolitan area, USA [17]. Their study confirmed that route length and slopes do have a negative effect on cyclists' route choice. In addition, they found that high traffic volumes, high turn frequencies, and traffic signals are also repellant road attributes according to the cyclists' route choice. Zimmermann et al. estimated a link-based bike route choice model from a sample of GPS observations (103 users and 648 trips) in the city of Eugene, Oregon [18]. Their study confirmed the sensibility of cyclists to distance, traffic volume, slopes, crossings and the presence of bike facilities, distinguishing between average slope above or below 4%, and traffic volume above or below 8000 vehicles per day. More recently, Bernardi et al. analyzed the GPS traces recorded by approximately 280 bicycle users throughout the Netherlands [19]; the results show a high usage of cycleway links and the preference of the shortest route by frequent cyclists. Casello and Usyukov used GPS data on cyclists' activities to estimate a generalized-cost function that reflects the cyclists' evaluation of path alternatives [20]: their model correctly predicted the revealed path choice for 65% of the examined trips.

One of the main problems with GPS data is its representativeness, because data collection is usually provided on a volunteer basis, which is not necessarily representative for the entire population [21]. This type of problem has been highlighted by Jestico et al. [22], who used data provided by strava.com to quantify how well crowdsourced fitness app data represent ridership through a comparison with manual cycling counts in Victoria, British Columbia. Another problem is the level of detail of the network: in many cases, the success of identifying the correct network links from GPS points is limited if the bike network model is not sufficiently detailed [23,24].

This paper explains how to estimate the city-wide bicycle flows and how to identify weak points of the road network in terms of bicycle friendliness. Both methods are data driven, explicit and do not require the calibration of sophisticated models. Nevertheless, a route choice model is also calibrated explaining the reasons for deviations at certain road links.

The paper is organized as follows. Section 2 describes the study area and the features of the bike network. Section 3 depicts the bicycle flows obtained by traditional (manual and instrumental) counting methods and by GPS data collected by smartphone application. Section 3 identifies a correlation between cycling counts and GPS data and describes the bicycle flow reconstruction method. In Section 4, a deviation analysis is carried out and a Logit model is calibrated in order to shed more light on the reasons for the decision of individuals to accept deviations by assessing the route characteristics of chosen routes and shortest routes. Section 5 discusses the results of the analysis. Concluding remarks and future research directions are presented in Section 6.

2. Dataset Description

2.1. Study Area

Bologna is a northern Italian city with approximately 390,000 inhabitants [25]. The climate is convenient for cycling all year, with an annual average temperature slightly below 15 °C and low rainfall (about 700 mm rain/year and 74 days of rain per year). Figure 1 shows an overview map to facilitate the location of the city of Bologna, including a zoom on the city center (see box in Figure 1 bottom left).

The home-to-work bicycle mode share was 8.2% in 2011 [26], which is relatively high compared with other medium to large Italian cities [27]. Nevertheless, the car ownership equals 0.515 cars per inhabitant [25], which corresponds to 0.97 cars per household. Based on a survey carried out by TNS opinion & social network in the 28 member states of the European Union between the 11th and 20th of October 2014, the average bicycle mode share was 8.0% [28], whereas in Italy the percentage of people who frequently commute by bike was approximately 4.7% in 2017 [27]. In addition, all major Italian municipalities show an average car ownership of 0.616 per person and a motorcycle ownership of 0.132 per person [29].

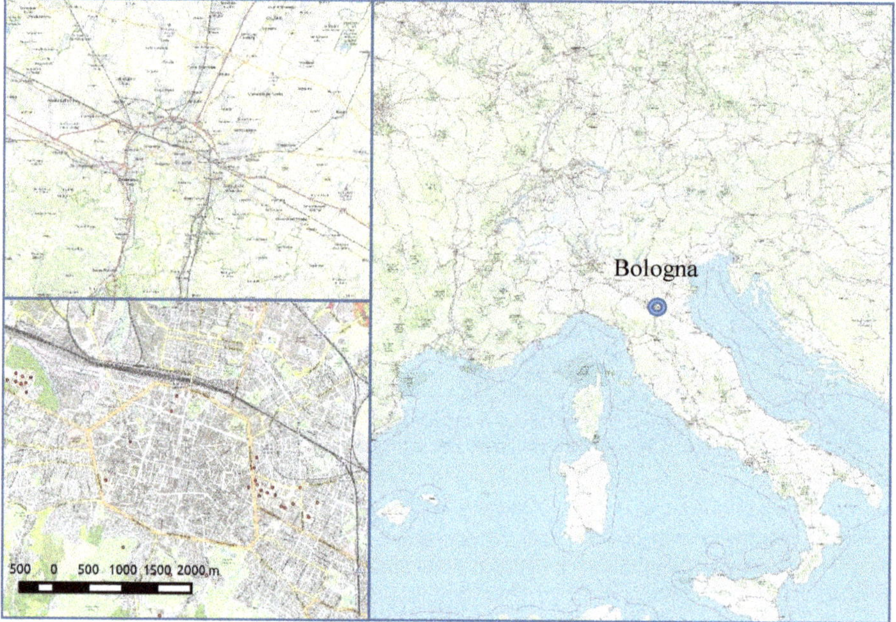

Figure 1. Bologna location in the Italian contest.

2.2. Bicycle Network

The municipality of Bologna has made substantial investments in bikeways during the past decade and to date the city offers 129 km bikeways of different types: exclusive access and mixed access with pedestrians or buses [25]. The bicycle network layout is composed of 13 main radial bicycle paths connecting the suburbs to the city center and many other bikeways connecting the radial bike-paths. The bikeway meters per citizen increased by 45% starting with 0.228 m/citizen in the year 2009 and reaching 0.330 m/citizen in 2018 [25]. This is an almost linearly-increasing expansion of the cycling infrastructure. The bike-network map is illustrated in Figure 2.

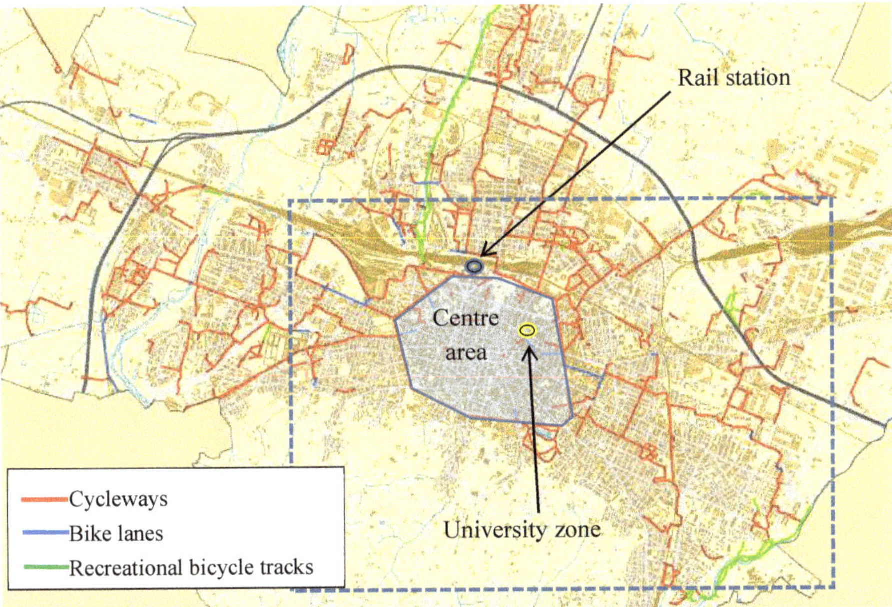

Figure 2. The bike-network map of Bologna [24] and study area (dashed lines).

3. Bicycle Flow Analysis

3.1. Cyclists' Flows from Traditional Counting Methods

During the period 2009–2018, manual and instrumental counts of cyclists were carried out by DICAM-Transport of the University of Bologna [30]: the bicycle counts were conducted from September to October of each year within the study area as shown in Figure 2. In recent years, counting has also been performed in May with the aim to evaluate the difference in bicycle flows between different periods of the same year. The locations of bicycle counters have been selected adopting representative and targeted locations: the sites include different geographic areas of the city, different types of bikeways, as well as "pinch points" (i.e., locations where cyclists must converge to cross a barrier) [10]. The 46 (bidirectional) road-sections monitored in 2018 are showed in Figure 3, highlighting the spatial distribution of measurement points. The monitored road-sections included the 13 main radial bicycle paths. Figure 3 also includes images of different typical bikeway types in Bologna.

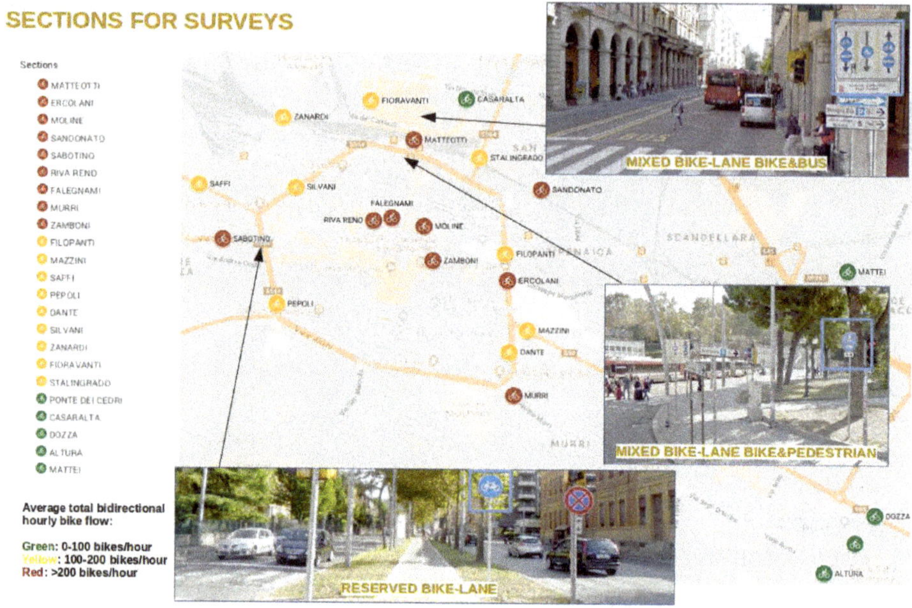

Figure 3. Road sections monitored in 2018—Legend shows sections sorted by flow value.

Manual and instrumental counting was conducted at each road section from 08:30 to 10:30 on weekdays. The trips purpose during this time period is most likely commute trips for the purpose of "work" or "study". It is further assumed that commute trips have a clear destination, with a low occurrence of round-trips or random trips for recreational purposes. The total average flows increased between 2009 and 2018 by approximately 75%, which is significantly greater than the increase in bikeway meters per inhabitant in the same period.

Figure 4 shows the correlation between bikeway meters per inhabitant and the total average bicycle flows: each point represents one year from 2009 to 2018.

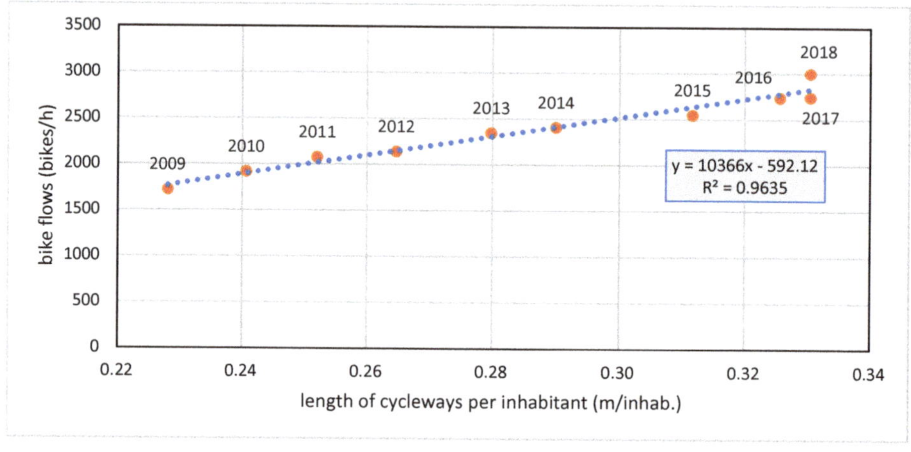

Figure 4. Regression function between length of cycleways per inhabitant and bike flows.

As shown in Figure 4, the total average bike flows are positively and highly correlated with the length of cycleways per inhabitant ($R^2 = 0.96$). In the city of Bologna, people use bicycles more often than in the past. Surely, such an increase in cycling is determined, like in other cities, by an integrated package of many different and complementary measures, including infrastructure provision, pro-bicycle programs, supportive land-use planning and restrictions of car use [2]. However, today's bicycle network of Bologna connects the most popular origins and destinations, and the expansion of the cycling network has resulted in an increased level of safety as demonstrated by accident statistics [25]. The increasing bicycle use is also related to an increasing bicycle use of females, growing from a share of below 30% in 2009 to a share of 44% in 2018 [30].

Using the regression function from Figure 4, one can estimate that one additional centimeter of bikeway per inhabitant increases the average bicycle flow by approximately 100 cyclists per hour on the main sections of Bologna's bicycle network. Based on the length increase of the bicycle network, the estimated bicycle mode share is currently almost 10%, following the model proposed by Schweizer and Rupi [31]: their model describes the significant linear relationship between meters of cycling infrastructure per inhabitant and bike mode share ($R^2 = 0.81$), based on approximately 9000 questionnaires carried out in 14 cities in Central Europe.

3.2. Map Matched Cyclists' Volumes

A database with GPS traces has been obtained from a data collection initiative called the "European Cycling Challenge" (ECC) [32] which took place in May 2016. In particular, the city of Bologna participated in this initiative among other 51 cities from 18 European countries. In Bologna, 1123 participants, equal to 0.3% of the population, recorded the GPS traces of their bicycle trips during the month of May 2016 by means of a mobile phone application. Participation was on a voluntary basis. The total distance travelled by all participating cyclists was almost 200,000 km and the database contains over 7,998,000 GPS points, with 27,348 individual trips covering the entire road network of Bologna [32] (see Figure 5). There is an area in the southern part of Bologna (encircled in green on Figure 5), with a particularly low density of GPS points, most likely due to the mountains and gardens with bike paths in which the observed bicyclist activity is almost completely absent.

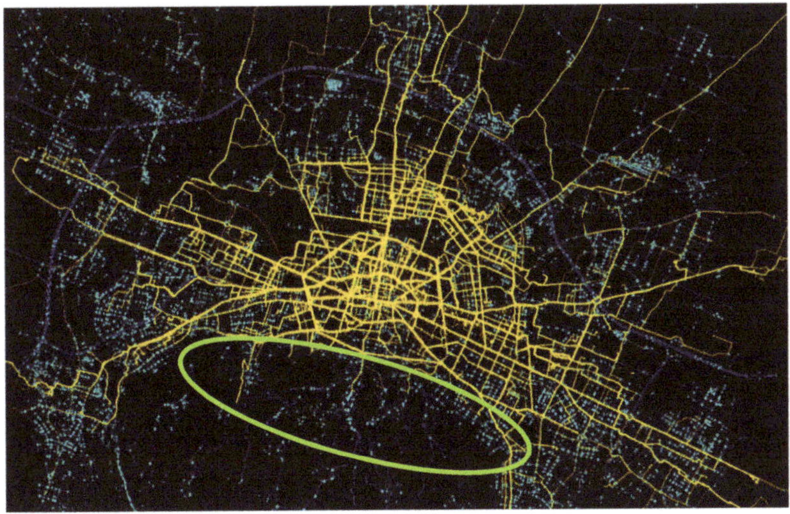

Figure 5. ECC 2016: observed cyclist activity in the Bologna network.

The present analysis focuses only on morning trips from 08:30 to 10:30 during work-days in order to be compatible with the manual and instrumental bicycle counts. During this period, 847 trips were recorded, of which 42% were female and the average age was 38 years. The share of trips carried out by workers with respect to students and the users' gender are very similar to the last trips census survey of Bologna [33]. However, the census is referred to the active people that use all means of transport. In addition, the share of GPS traces recorded by females is very similar to the share of females observed during the manual counts. Consequently, the sample of cyclists recording the GPS traces is representative of the gender of the counted cyclists. Unfortunately, the ECC database contains no information concerning trip purposes.

In order to obtain bicycle flows on network links, the GPS data has been matched to the road-network based on open street map (OSM). The OSM data has been extracted for the Bologna metropolitan area and converted into a SUMO (Simulation of Urban Mobility) network [34] using a software extension called SUMOPy [35] as reported in Rupi and Schweizer [23]. The SUMO network has been manually corrected and enhanced, such that cyclists could potentially pass everywhere, including footpaths and the opposite direction of one-way roads (which is an illegal behavior in Italy). The final network contains 13,959 nodes and 38,324 links. The employed map matching algorithm is part of SUMOPy and based on a method proposed by Marchal et al. [36] and improved by Schweizer et al. [37]. In order to match the GPS points to network links with a high accuracy and to obtain a large number of correctly matched GPS traces, the entire map-matching analysis consists of four phases [23]: (i) an initial filtering process, (ii) the actual map matching process, (iii) a post-filtering process, and (iv) a final analysis of the matched routes. Initially, many GPS traces could not be matched to the network due to missing links or missing access. Successively, the reasons for the failed matching of trips have been analyzed in detail and missing network links or road access attributes have been added. Finally, the map-matching process has been repeated with an increased number of successfully matched trips.

After the map-matching process and a filtering process ensuring a low error rate, 4029 map-matched routes, collected from 842 users, have been used. These traces correspond to 91.6% of all traces recorded during the considered morning period. It is worth mentioning that this percentage is significantly higher than that reported in other studies [23,24]. Starting from these map-matched routes, the bicycle flows (as the number of cyclists passing through each network link per hour) have been evaluated.

3.3. Estimated Cyclists' Volumes

A linear regression between the cyclists counted with traditional methods and the number of map matched GPS traces with links overlapping the monitored road sections has been carried out. The map matched bicycle volumes have been multiplied by a coefficient c in order to minimize the difference between the measured flows and flows derived from GPS data.

The regression, shown in Figure 6, is based on the flow-comparison at 23 monitored sections ($c = 0.91$).

The slope of the linear regression function is almost equal to one, highlighting that the average of map-matched cyclist volumes are equal to the average of manually counted cyclist volumes.

The relatively high level of correlation between the measured flows and the flows from the map matched GPS traces is evident.

Given the significant correlation between the GPS dataset and traditional counts, the linear relation between both flow types has been used to determine the flows on all network links where GPS points have been detected. The resulting link flows in cyclists per hour per direction are shown in the Figure 7. This map is particularly useful to quantify the spatial distribution of ridership and provide important cycling exposure data for safety studies. Starting from this map, it is possible to obtain the OD matrix of cyclists, the chosen routes, and the bicycle flow on every link of the network. This is essential information for modelling the cyclists' route choice behavior and for planning the bicycle network.

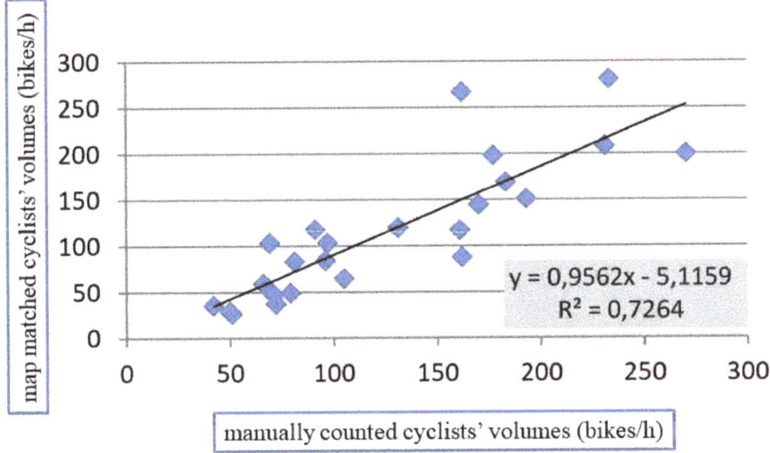

Figure 6. Regression function between manually counted cyclists' volumes and map matched cyclists' volumes (May 2016).

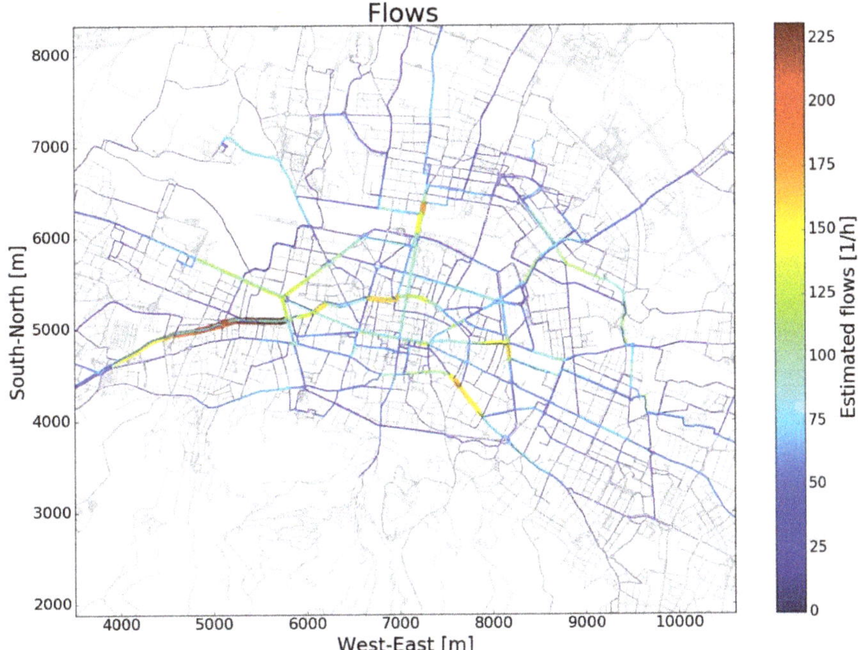

Figure 7. Estimated unidirectional bicycle flows in cyclists per hour during workday morning peak hours (from 8:30 to 10:30). Flows only on network links where GPS points have been detected.

In addition, Strava provides the cyclist heatmap of all cities in the world for trips using the Strava app. The heatmap is calculated by counting and normalizing the number of lines connecting recorded GPS points [38]. The Strava app collects mainly recreational trips and in particular sport trips. The Strava density heatmap of recorded GPS points in Bologna is reported in Figure 8. This figure highlights how the recorded trips are also spread in the south of Bologna, in mountain routes as well as in gardens provided with bike paths (encircled in green). Instead, ECC's traces from 8:30 to 10:30

cover the main cycle ways that directly connect different parts of city, but there is an absence of trips in the south part, compared with the encircled area of Figure 5. This is probably due to the difference in trip purposes, supporting the hypothesis that the ECC sample contains few leisure trips.

Figure 8. The Strava heat-map of Bologna.

4. Deviation Analysis

The deviation analysis aims to identify the network links which are the most avoided by all cyclists who registered GPS traces. The analysis starts with the following basic assumption: given the choice of two routes with identical properties (same safety, pavement, environment, etc.), cyclists would always choose the shortest one. If this is true, the cyclist would only accept a longer route if it offered better properties (safer, quieter, etc.). From a different perspective, if certain road links are avoided by deviating on alternative links, then the avoided links are supposed to possess fewer attractive characteristics with respect to the alternative, even though these characteristics may be good in the absolute sense. In an ideal bicycle network, no cyclists should feel constrained to take a longer route due to some repellant characteristics of the shortest route, or due to the better characteristics of longer routes. The most "avoided links" of the city's road network are therefore identified with the km of deviation caused to cyclists. The total deviation metric DM_i for each road link i is calculated in the following way:

1. For each matched route R_j of the set of all matched routes J, determine the shortest route S_j connecting the first and last link of each matched route.
2. For each matched route R_j, identify all K_j non-overlapping sections where links deviate from the shortest route. Set DR_{jk} contains all chosen links of the partial deviation k of route j and set SR_{jk} contains all links on the shortest route of deviation k and route j, as illustrated in Figure 9.
3. For each of these non-overlapping sections, calculate the partial deviation d_{jk} which is the difference between the length of the part of the chosen route segment DR_{jk} and the length of the corresponding part of the shortest route segment SR_{jk}; finally the deviation metrix DM_i of link i is the sum of partial deviations of all routes that contain link i on one of the shortest route segments. Analytically, the total deviation metric DM_i of a road link i is the sum of all partial deviations received from all non-overlapping sections of all matched trips and can be expressed as:

$$DM_i = \sum_{j \in J} \sum_{k=1}^{K_j} \delta_{ijk} \cdot d_{jk} \tag{1}$$

where $\delta_{ijk} = 1$ if SR_{jk} contains link i, otherwise 0. Let L_i be the length of link i; then the partial deviation is given by $d_{jk} = \sum_{i \in DR_{jk}} L_i - \sum_{i \in SR_{jk}} L_i$.

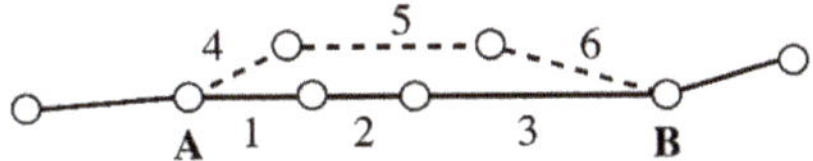

Figure 9. Illustration of the calculation of the total deviation metric for the non-overlapping route section between nodes A and B.

Figure 9 shows links 1, 2 and 3 which are not chosen, despite they are part of the shortest route (solid line); whereas, links 4, 5 and 6 are part of the chosen route (dashed line). In case of the non-overlapping section between node A and B shown in Figure 9, the chosen route DR_{jk} is constituted by links 4, 5 and 6, while the shortest route section SR_{jk} contains links 1, 2 and 3. The partial deviation d_{jk} of links in SR_{jk} equals to $d_{jk} = L_4 + L_5 + L_6 - (L_1 + L_2 + L_3)$. The total deviation metric for the central part of Bologna network is shown in Figure 10.

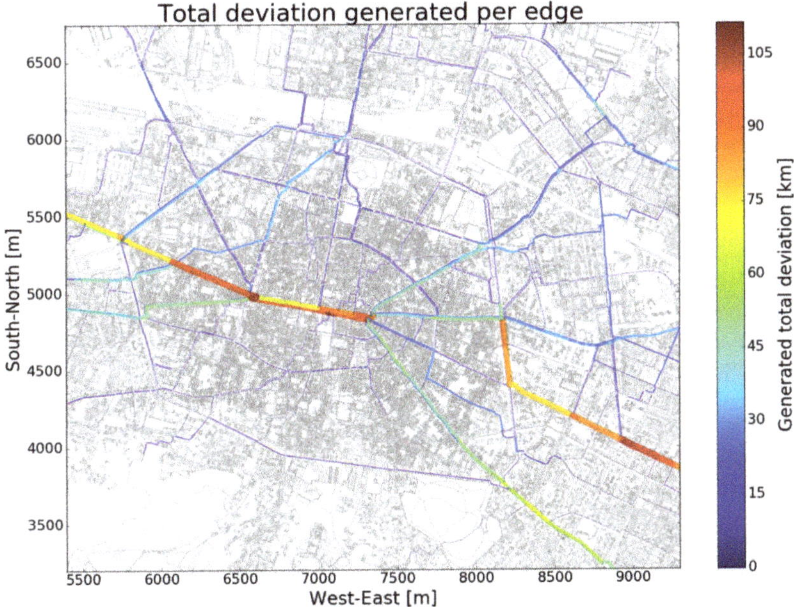

Figure 10. Total deviation metric determined for the central part of Bologna network.

The highest total deviation metric can been seen on the main radial roads from and into the city center. As seen in Figure 10, these are also roads with high bicycle flows. This means that many cyclists actually do use these radial roads but also many try to avoid them. Note that there are also roads in the city center with high bicycle flows, but generating almost no deviations. For a discussion of these findings, see Section 5.

On average, the chosen route parts are 20% longer with respect to the shortest route parts. Analyzing the road attributes of the chosen part and the shortest part of all non-overlapping sections of all trips, the causes for the deviations become clearer—see the first three columns of Table 1. As expected, cyclists accept deviations in order to travel on roads with: (1) a high share of reserved bikeways, (2) a high share of low priority roads (roads with one lane per direction and speed limits of

30 km/h), (3) a low intersection density, and (4) a low share of mixed access, such as lanes with bike/bus access or lanes where bikes and pedestrians are allowed. This last result confirms the findings of the research carried out by Bernardi et al. [39], in which the authors quantified the effects and frequencies of disturbances on bicycle facilities, particularly from pedestrians and buses.

Table 1. Road link attributes of chosen and shortest routes of non-overlapping sections and on overlapping sections.

	Non-overlapping Sections			Overlapping Sections
	Shortest route	*Chosen route*	*Chosen vs Shortest*	*Chosen and shortest route*
Total length [km]	8265	9975	+20.7%	7130
Mixed road access share	32.6%	25.7%	−21.2%	28.4%
Low priority road share	50.0%	74.1%	+48.2%	40.2%
Reserved bikeway share	16.4%	39.2%	+139.0%	23.5%
Intersection density [1/km]	18.5	15.9	−14.2%	16.1

The statistics of the road link attributes of the overlapping sections of each trip (i.e., all links where the chosen and shortest routes coincide) are presented in the last column of Table 1. It becomes evident that the values of the mixed road access share, the reserved bikeway share and the intersection density are in between the values of the shortest route (column 1) and the chosen route (column 2) of the non-overlapping sections. One could conclude that cyclists tend to deviate if road attribute values are below/above those of the overlapping sections. An exception is the low priority road share, where the overlapping sections show values even below that of the shortest route.

In order to shed more light on the decision of individuals to accept deviations, a Logit model is calibrated, where the user has the choice between two alternatives of non-overlapping route segments (as illustrated in Figure 9), where one of the alternatives is the shortest route. The systematic utility function V_i of alternative i is defined as:

$$V_i = \beta_1 D_i + \beta_2 B_i + \beta_3 LP_i \qquad (2)$$

where D_i is the distance of the route segment, B_i is the share of exclusive bikeway, and LP_i is the share of low priority roads as percentage of the respective distance. The set of observations has been prepared as follows. In a first subset, the route sections have been considered, where the chosen route is different from the shortest route. In a second subset, route sections have been identified, where the shortest route completely coincides with the chosen route. In this case, a longer route alternative (that has not been chosen by the cyclist) has been generated as follows: the second shortest route that connects the extremities of the shortest route section is determined such that the second shortest route section does not overlap with the shortest route section. This is similar to the method applied by Marchal et al. [36]. In this way, a route alternative is generated that is the closest possible to the chosen (and shortest) route alternative. In order to avoid a bias towards longer or shorter distances, the size of the first and second observation subset are kept equal.

The calibration resuls of a total of 4678 observations is shown in Table 2. The attributes chosen are all significant and $R^2 = 0.160$. The small parameter values result in Odds ratios close to one, which is reasonable considering that attribute values are in the order of 10^{-2}–10^{-3}. Other attributes like the node density or the share of mixed bikeway access have turned out not to be significant when included in this model. The signs of the model parameters are reasonable—see also the discussion in Section 6. The calibration has been repeated with GPS traces in Bologna from the ECC of the year 2015. The result of this calibration shows parameter values within the standard error bounds of the result from ECC of the year 2016 shown in Table 2.

Table 2. Calibration results of Logit model from Equation (2).

Attribute	Parameters β_j	Odds Ratio	Std. err	z	$p > |z|$
Distance D_i	−0.0007	0.9992	$9.10 \ 10^{-5}$	−7.4585	$8.7531 \ 10^{-14}$
Reserved bikeway share B_i	0.0228	1.0230	$8.82 \ 10^{-4}$	23.1713	$8.8673 \ 10^{-119}$
Low priority road share LP_i	−0.0085	0.9915	$6.88 \ 10^{-4}$	−12.0599	$1.7192 \ 10^{-33}$

One can use Equation (2) to estimate the deviation necessary to equilibrate the systematic utilities of both route alternatives. Setting $V_1 = V_2$, and resolving for the deviation yields in

$$D_1 - D_2 = \frac{\beta_2}{\beta_1}(B_2 - B_1) + \frac{\beta_3}{\beta_1}(LP_2 - LP_1), \tag{3}$$

which is the difference in distance, depending on the difference in exclusive bikeway share and the difference in low priority road share. The deviation $D_1 - D_2$ obtained from Equation (3) ensures a path choice probability of 50%.

5. Discussion

Regarding the estimation of all bicycle flows on the network, the high agreement of flows from GPS traces and manual/instrumental counts ($R^2 = 0.73$) is significantly better than the results obtained by previous studies, e.g., Jestico et al. [22] obtained an R^2 of 0.4 for the a.m. peak period. The reason for this difference is likely due to the more detailed network model of Bologna, representing better the cyclists' freedom to move on all possible links in both directions. Based on this correlation, one crowdsourced cyclist corresponds in average to 59 cyclists counted with traditional methods, which is consistent with previous findings in [22].

Although crowdsourced cyclists represent a small portion of all cyclists, the flows obtained from the map matched GPS data are consistent with the observed flows on the main sections of the Bologna cycle network. Only a few outliers emerge, most likely due to two potential error sources: (1) the days of data sampling of the two methods do not entirely overlap because the GPS records have been registered during the whole month of May, while the traffic counts have been conducted only for two weeks of the same month; (2) the sampling hours do not exactly overlap either, because the GPS traces are selected by their begin time while the traditional methods count the cyclists actually passing by the road-section during the analyzed time interval.

The applied deviation metric from Section 4 quantifies the total deviations of cyclists generated by single road links, but the metric itself does not identify the reasons for the deviations. However, it becomes evident that those radial roads with high bike flows and high deviations in Figure 10 are characterized by an absence of reserved bike lanes, a high level of bus traffic, often on reserved bus lanes, and a high density of intersections. In contrast, those roads where bicycle flows are high but deviations are low, are characterized by low motorized traffic volumes, a high share of bike lanes and the absence of major bus routes.

Nevertheless, the total deviation metric depends on the presence of alternative routes with respect to the shortest route and their respective road attributes: in case there are no feasible route alternatives to avoid a certain link, then the total deviation metric of the respective link is zero, even though the attributes may be unfavourable. In case the shortest route has favourable link attributes but the alternative has even more favourable link attributes, then the total deviation metric is high despite the good conditions on the shortest route. The former case is the most severe as criticalities of unfavourable roads for cyclists without route alternatives remain undiscovered by the deviation analysis.

The calibration of a binominal logit model allows to determine the choice probability, given the attributes of two non-overlapping route alternatives. The attributes distance, share of exclusive bikeways and share of low-priority roads have been found to be significant. The negative sign of β_1 and the positive sign of β_2 are expected as a longer distance is a disincentive and the presence of a an exclusive bikeway is an incentive for cyclists. The negative sign of β_3 related to the share of low-priority roads seems to contradict the previous analyses where the average route on deviations contains a lower share of low priority roads with respect to routes where shortest and chosen routes overlap. One explanation could be the way the alternative routes are generated: probably the second shortest route between two points does use minor roads which cyclists would typically avoid, or cyclists are simply not aware of such alternatives. The determination of road priorities is a heuristic algorithm derived from OSM attributes such as speed limits, number of lanes and access restrictions. It is possible that some types of low priority roads are in reality not attractive to cyclists. A possibility to avoid such problems would be to consider only alternative routes used by at least one cyclist instead of using the adopted route generation method. Another solution would be to specify a model which quantifies the probability to accept a route by considering only the link attributes of the chosen routes, enhanced by a dummy variable indicating whether the chosen route is also the shortest route. Such a model would definitively avoid the problem of alternative route generation. However, modelling errors could be introduced by ignoring all attributes of alternative routes.

The result from Equation (3) relates the differences in road attributes between two alternative routes with the distance that would compensate those differences. The meaning of this result shall be explained by a simple numerical example: assuming two route alternatives, one with a 100% exclusive bikeway and the other without any bikeway, and both alternatives are without priority roads. In such a case, the first alternative could be 3.2 km longer than the second while still attracting 50% of the cyclists.

6. Conclusions

In this research, the cyclists' flows obtained by traditional counting methods have been compared with GPS traces from smartphones at the same locations and during the same time period.

Although crowdsourced cyclists represent often a small portion of all cyclists, they do represent well the ridership of Bologna in terms of cyclists' volumes and gender distribution. This result emerges clearly by comparing traditional counting method with GPS traces, confirming their representativeness of the population. The correlation between cycling counts and GPS data collected by smartphones has been relatively high, with an R^2 value of 0.73. This correlation is significantly higher than the results obtained by other studies, most likely due to the more detailed representation of the Bologna network, including footpaths in parks and the possibility to cycle one-way roads in both directions. Due to this high correlation, it has been possible to estimate the absolute bicycle flows on all network links by an appropriate scaling of the map-matched flows. The cyclists' routes are of great value for the planning of cycling infrastructure and the drafting of cycling policies. The proposed method, which combines bicycle counts at a few main road sections with areas covering GPS traces, can readily be applied in other cities in order to reliably estimate the absolute bike flows of an entire urban area.

GPS data have been further used to determine the total deviation metric, which counts the total deviations that a road link causes to cyclists. The total deviation metric is useful to identify weak links of the cycling network, but it does not identify the reason why certain road links are avoided. However, applying the total deviation metric to the Bologna road network, the highest deviation has been seen on trafficked roads without physically protected bike lanes. Also, roads with reserved bus lanes, which are open for bicycles too, showed high deviation rates. Further analyses of chosen and shortest road sections have shown that cyclists are willing to make deviations when the alternative route provides a high share of reserved bikeways, a high share of low-priority lanes, a low intersection density and a low share of roads with mixed traffic (cyclists with buses and pedestrians). Planners should take the deviation metric into considerations for either bike-path construction or bike-network interventions. Obviously, the total deviation metric does not reveal deviations if there are no alternatives to avoid a certain road link. The map-matched traces allowed to calibrate a discrete choice model between two route alternatives, considering distance, share of exclusive bikeway and share of low-priority roads. A longer distance and a higher share of low-priority roads appear to decrease the choice probability, while a higher share of exclusive bikeways does increase the choice probability, as expected. With the same model, it has been possible to quantify the tolerated deviation length in function of the road attributes.

In future works, the representativeness of the results could be improved by statistically weighting the GPS traces according to different person attributes, such as occupation, gender, or age. The route choice model could be enriched by more significant attributes like traffic light density, junctions with left-turns, or junctions with side-roads entering from the right side. In particular, the low-priority road attribute needs to be further refined. The generation of longer route alternatives could be replaced by actually chosen routes and models using only attributes of the chosen routes shall be tested.

Author Contributions: Study conceptualization, Federico Rupi and Joerg Schweizer; Methodology, Federico Rupi and Joerg Schweizer; Programming and data retrieval, Federico Rupi, Joerg Schweizer and Cristian Poliziani; Formal GIS analysis, Joerg Schweizer and Cristian Poliziani; Writing—review and editing, Federico Rupi, Joerg Schweizer and Cristian Poliziani.

Funding: Open access funding provided by Department of Civil, Chemical, Environmental, and Materials Engineering (DICAM)—University of Bologna.

Acknowledgments: We are grateful to SRM Bologna srl for providing the GPS data of the European Cycling Challenge 2016.

Conflicts of Interest: The authors declare no conflict of interest.

References

1. Dill, J.; Carr, T. Bicycle Commuting facilities in major US cities: If you build them commuters will use them—Another look. *Transp. Res. Rec.* **2003**, *1828*, 116–123. [CrossRef]
2. Pucher, J.; Buehler, R. Making cycling irresistible: Lessons from The Netherlands, Denmark and Germany. *Transp. Rev.* **2008**, *28*, 495–528. [CrossRef]
3. Dill, J.; Voros, K. Factors Affecting Bicycling Demand: Initial Survey Findings from the Portland, Oregon, Region. *Transp. Res. Rec.* **2007**, *44*, 9–17. [CrossRef]
4. Stinson, M.; Bhat, C. Commuter Bicyclist Route Choice: Analysis using a Stated Preference Survey. *Transp. Res. Rec.* **2003**, *39*, 107–115. [CrossRef]
5. Winters, M.; Davidson, G.; Kao, D.; Teschke, K. Motivators and deterrents of bicycling: Comparing influences on decisions to ride. *Transportation* **2011**, *38*, 153–168. [CrossRef]
6. Sener, I.N.; Eluru, N.; Bhat, C.R. An analysis of bicycle route choice preferences in Texas, US. *Transportation* **2009**, *36*, 511–539. [CrossRef]
7. Howard, C.; Burns, E.K. Cycling to Work in Phoenix: Route Choice, Travel Behavior, and Commuter Characteristics. *Transp. Res. Rec.* **2001**, *1773*, 39–46. [CrossRef]
8. Nelson, A.C.; Allen, D. If you build them commuters will use them–Association between bicycle facilities and bicycle commuting. *Transp. Res. Rec.* **1997**, *1578*, 79–83. [CrossRef]
9. Niemeier, D.A. Longitudinal Analysis of Bicycle Count Variability: Results and Modeling Implications. *J. Transp. Eng.* **1996**, *122*, 200–206. [CrossRef]
10. Ryus, P.; Ferguson, E.; Laustsen, K.M.; Schneider, R.J.; Proulx, F.R.; Hull, T.; Miranda-Moreno, L. *Guidebook on Pedestrian and Bicycle Volume Data Collection*; National Cooperative Highway Research Program Report 797; The National Academies Press: Washington, DC, USA, 2014.
11. Griffin, G.; Nordback, K.; Götschi, T.; Stolz, E.; Kothuri, S. Monitoring bicyclist and pedestrian travel and behavior, current research and practice. Transportation research circular E-C183. 2014. Available online: http://www.trb.org/Publications/Blurbs/170452.aspx (accessed on 31 May 2019).
12. Pritchard, R.; Frøyen, Y.K.; Bernhard, S. Bicycle Level of Service for Route Choice—A GIS Evaluation of Four Existing Indicators with Empirical Data. *Int. J. Geo-Inf.* **2019**, *8*, 214. [CrossRef]
13. Griffin, G.P.; Jiao, J. Where does bicycling for health happen? Analyzing volunteered geographic information through place and plexus. *J. Transp. Health* **2015**, *2*, 238–247. [CrossRef]
14. Dill, J. Bicycling for Transportation and Health: The Role of Infrastructure. *J. Public Health Policy* **2009**, *30*, 95–110. [CrossRef] [PubMed]
15. Menghini, G.; Carrasco, N.; Schüssler, N.; Axhausen, K.W. Route choice of cyclists in Zurich. *Transp. Res. Part A* **2010**, *44*, 754–765. [CrossRef]

16. Hood, J.; Sall, E.; Charlton, B. A GPS-based bicycle route choice model for San Francisco, California. *Transp. Lett. Int. J. Transp. Res.* **2011**, *3*, 63–75. [CrossRef]

17. Broach, J.; Dill, J.; Gliebe, J. Where do cyclists ride? A path choice model developed with revealed preference GPS data. *Transp. Res. Part A* **2012**, *46*, 1730–1740.

18. Zimmermann, M.; Mai, T.; Frejinger, E. Bike route choice modeling using GPS data without choice sets of paths. *Transp. Res. Part C* **2017**, *75*, 183–196. [CrossRef]

19. Bernardi, S.; La Paix-Puello, L.; Geurs, K. Modelling route choice of Dutch cyclists using smartphone data. *J. Transp. Land Use* **2018**, *11*, 883–900. [CrossRef]

20. Casello, J.M.; Usyukov, V. Modeling cyclists' route choice based on GPS data. *Transp. Res. Rec.* **2014**, *2430*, 155–161. [CrossRef]

21. Watkins, K.; Ammanamanchi, R.; LaMondia, V.; Le Dantec, C.A. Comparison of Smartphone-based Cyclists GPS Data Sources. In Proceedings of the Transportation Research Board—95th Annual Meeting, Transportation Research Board of the National Academies, Washington, DC, USA, 10–14 January 2016.

22. Jestico, B.; Nelson, T.; Winters, M. Mapping ridership using crowdsourced cycling data. *J. Transp. Geogr.* **2016**, *52*, 90–97. [CrossRef]

23. Rupi, F.; Schweizer, J. Evaluating cyclist patterns using GPS data from smartphones. *ITE Intell. Trans. Syst.* **2018**, *12*, 279–285. [CrossRef]

24. Khatri, R.; Cherry, C.R.; Nambisan, S.S.; Han, L.D. Modeling Route Choice of Utilitarian Bikeshare Users with GPS Data. *Transp. Res. Rec.* **2016**, *2587*, 141–149. [CrossRef]

25. Municipality of Bologna, Statistics. Available online: https://www.comune.bologna.it/iperbole/piancont/dati.html (accessed on 1 May 2019).

26. ISTAT Statistics. 2011. Available online: https://www.istat.it/it/censimenti-permanenti/censimenti-precedenti/popolazione-e-abitazioni/popolazione-2011 (accessed on 1 May 2019).

27. Istat Spostamenti Quotidiani e Nuove Forme di Mobilità. 2018. Available online: istat.it/it/files//2018/11/Report-mobilità-sostenibile.pdf (accessed on 1 May 2019).

28. Eurobarometer. Special Eurobarometer 422a, Quality of Transport. 2014. Available online: ec.europa.eu/public_opinion/archives/ebs/ebs_422a_en.pdf (accessed on 1 May 2019).

29. Conto Nazionale dei Trasporti. 2018. Available online: http://www.mit.gov.it/comunicazione/news/conto-nazionale/online-il-conto-nazionale-delle-infrastrutture-e-dei-trasporti (accessed on 1 May 2019).

30. Municipality of Bologna Rilevamento dei Flussi di Biciclette sulle Principali Piste Ciclabili Presenti nel Territorio del Comune di Bologna. Elaborazione dei Dati Raccolti e Confronto con le Serie Storiche Disponibili. 2018. Available online: www.comune.bologna.it/media/files/report_flussi_ciclabili_2018.pdf (accessed on 1 May 2019).

31. Schweizer, J.; Rupi, F. Performance evaluation of extreme bicycle scenarios. *Procedia-Soc. Behav. Sci.* **2014**, *111*, 508–517. [CrossRef]

32. European Cycling Challenge. 2016. Available online: www.europeancyclingchalleg.org (accessed on 1 May 2019).

33. Statistical Analysis of Bologna Trips. Available online: https://www.comune.bologna.it/iperbole/piancont/Cens_Pop_2011/pendolarismo/Pendolarismo.pdf (accessed on 1 May 2019).

34. Sumopy. Available online: https://sumo.dlr.de/wiki/Contributed/SUMOPy (accessed on 1 May 2019).

35. SUMO. Available online: http://sumo.sourceforge.net/userdoc/ (accessed on 1 May 2019).

36. Marchal, F.; Hackney, J.K.; Axhausen, K.W. Efficient map matching of large Global Positioning System data sets: Test. on speed-monitoring experiment in Zurich. *Transp. Res. Rec.* **2005**, *1935*, 93–100. [CrossRef]

37. Schweizer, J.; Bernardi, S.; Rupi, F. Map-matching algorithm applied to bicycle global positioning system traces in Bologna. *ITE Intell. Transp. Syst.* **2016**, *10*, 244–250. [CrossRef]

38. Strava Global Heatmap. Available online: https://www.strava.com/heatmap#13.41/11.34182/44.48590/hot/all (accessed on 1 May 2019).

39. Bernardi, S.; Krizek, K.J.; Rupi, F. Quantifying the role of disturbances and speeds on separated bicycle facilities. *J. Trans. Land Use* **2016**, *9*, 105–119. [CrossRef]

 International Journal of *Geo-Information*

Article

An Agent-based Model Simulation of Human Mobility Based on Mobile Phone Data: How Commuting Relates to Congestion

Hao Wu [1], Lingbo Liu [2], Yang Yu [2,*], Zhenghong Peng [1], Hongzan Jiao [2] and Qiang Niu [2]

[1] Department of Graphics and Digital Technology, School of Urban Design, Wuhan University, Wuhan 430072, China
[2] Department of Urban Planning, School of Urban Design, Wuhan University, Wuhan 430072, China
* Correspondence: yuyang1@whu.edu.cn; Tel.: +86-27-6877-3062

Received: 5 June 2019; Accepted: 20 July 2019; Published: 23 July 2019

Abstract: The commute of residents in a big city often brings tidal traffic pressure or congestions. Understanding the causes behind this phenomenon is of great significance for urban space optimization. Various spatial big data make the fine description of urban residents' travel behaviors possible, and bring new approaches to related studies. The present study focuses on two aspects: one is to obtain relatively accurate features of commuting behaviors by using mobile phone data, and the other is to simulate commuting behaviors of residents through the agent-based model and inducing backward the causes of congestion. Taking the Baishazhou area of Wuhan, a local area of a mega city in China, as a case study, we simulated the travel behaviors of commuters: the spatial context of the model is set up using the existing urban road network and by dividing the area into space units. Then, using the mobile phone call detail records of a month, statistics of residents' travel during the four time slots in working day mornings are acquired and then used to generate the Origin-Destination matrix of travels at different time slots, and the data are imported into the model for simulation. Under the preset rules of congestion, the agent-based model can effectively simulate the traffic conditions of each traffic intersection, and can induce backward the causes of traffic congestion using the simulation results and the Origin-Destination matrix. Finally, the model is used for the evaluation of road network optimization, which shows evident effects of the optimizing measures adopted in relieving congestion, and thus also proves the value of this method in urban studies.

Keywords: mobile phone data; residents commuting behavior; agent-based model; urban planning; traffic congestion

1. Introduction

With the expansion of urban scale and the rapid growth of motor vehicles, traffic congestion has become an increasingly serious urban problem, and the tidal traffic generated by the commuting of residents is believed to be one of the major causes of traffic congestion [1,2]. Traffic congestion not only brings energy waste and environmental pollution [3], but is also believed to negatively affect public health [4]. To address this problem, common approaches including economic or policy measures based on econometric models are used, such as congestion pricing [1,5–7], encouraging the use of public transport [8–10], etc. However, models commonly used in these approaches are static, in which residents' distribution and mobility in space are seldom considered, thus bringing inaccuracy of results [11]. Currently, increasing the availability and utilization of urban spatial big data, especially location-based service (LBS) data from GPS devices, smart cards, and mobile phones, make it possible to describe urban residents' travels more accurately on a finer scale [12]. Data of residents' mobility over time and space can be used for urban geographic mapping [13], epidemiological analysis [14,15], real-time urban monitoring [16], etc. and can also be used for recognition of urban spatial features [17–19] or measurement of urban vibrancy [20]. Another important scenario of application is the study of residents' commuting and urban transport, including the identification of commuting areas and commuting distances [21,22], and the acquisition of commuter Origin-Destination (OD) matrices [23–25]. Among various sources of location data, mobile phone data have been widely employed in studies such as residents' commuting thanks to its extensive coverage, passive data collection, and the fact that its data acquisition requires no extra equipment. In comparison, alternative data sources, such as smart card data or taxi GPS data, have equivalent difficulties in data coverage but much smaller population coverage [26]. Results of studies based on new data have been shown to have higher accuracy compared to those that are based on statistical data or measured data, proving the effectiveness of big data application in urban studies. In general, most of these studies are at an early stage of describing urban phenomena through data, few studies attempt to go further such as using big data to identify the connection between residents' travel and traffic congestion, or to predict and evaluate measures for traffic improvement [27,28].

After obtaining relatively accurate data for residents' commuting, modeling and simulation can be an approach to identifying the mechanism and rules behind the functions of urban spaces. The agent-based model (ABM) is considered one of the most effective techniques for simulating complex systems and thus has great advantage to study cities, which are typically complex systems [29]. The distributed characteristics of ABM enable it to reflect differences in the behaviors of different types of individuals [30]. Therefore, the model can be used to simulate traffic flow and residents' behaviors in urban transport, including the residents' choice of travel modes [31] and carpooling models [32]. Other research applications are found in the optimization of bus routes [33,34], the simulation of the functioning of the urban composite transportation system [35], the evaluation of the impact of the intercity high-speed railway on the ecological environment [36], etc. From the perspective of development trends in research, studies are moving from the simulation of individual decision-making to that of the composite flow of urban traffic, with increasing complexity of simulation. However, most of these simulations are still based on survey data, which not only are expensive and time-consuming to acquire but also lack details of residents' travel behaviors. For these reasons, big data are considered to be a better data source for studies of residents' travel behaviors [37].

Among others, call detail records (CDRs) are commonly employed as a kind of urban spatial big data. Compared with traditional survey data, it has higher sample coverage, time efficiency in acquisition, and higher time resolution [38]. On the other hand, as an effective tool for studying urban spaces, ABM has long been fettered by the lack of data in its earlier developments and it sees great potential in the current context of smart city development [39]. Therefore, using CDRs in ABM offers great promises for traffic simulation that reflects actual urban spatial environment and the spatial distribution of residents. The simulation, in turn, can be used to analyze the causes of traffic congestion and even to predict traffic conditions under different application scenarios. In previous studies, although big data is gradually applied to generate the OD matrix of the residents' travels and to predict traffic pressure in the actual urban road network, the following weaknesses still exist: first, most of these studies were conducted on a macro scale of the whole city. At such a scale, road capacity is often ignored, despite the fact that it is crucial for relieving traffic congestion; second, most studies presume that all commuting travels begin simultaneous, without addressing the differences in traffic volume in different time periods. Apparently, considerable errors may occur in the prediction of traffic conditions on the micro scale. Therefore, these two points were taken into consideration in the model employed in the present study.

2. Materials and Methods

The present research comprises two major parts: the acquisition of the features of residents' commuting behavior and simulation of commuting behavior of urban residents.

2.1. Mobile Phone Data Processing

As mobile phone data is directly related to the spatial distribution of the base stations, its accuracy in positioning is also determined by the density of base stations and varies across different areas. In addition, due to the different way of work and life of various users, the acquired phone call behavior is also fuzzy data with an uneven distribution over time. In general, mobility studies using mobile phone data usually take areas with densely distributed base stations such as city centers as case studies. User's location is represented by the location of the base station that has recorded the most frequent phone calls by the user within a specific period (one month or several months) at a specific time (working hours or at an interval of several hours). Subsequently, by associating the locations of various base stations along the timeline, user's mobility trajectory can be generated using CDR data [17,19,21,23–25,38,40–42]. The present study uses a similar approach while focusing on the rush hours and dividing the timeline on an hourly basis. Data are processed in combination with the ArcGIS platform, and the raw mobile phone data (Table 1) comprise CDRs of 7 million users in the case city over a time period of one month.

Data processing followed the procedures below:

a. Invalid data and users who make less than three phone calls per month were removed. After the screening, 3.8 million users remained.

b. Users' CDR data were sorted into working hours (7:00 to 18:00, Monday to Friday) and non-working hours (Saturday and Sunday, and 7pm to 6am on weekdays). Base station locations with the highest call frequency during the two time periods were identified as the place of work and residence, respectively.

c. Frequency of phone calls on working days (Monday to Friday) was calculated based on user's CDRs every 24 hours, and field value of the base station ID with the highest call frequency during the period was extracted (as shown in Table 2).

d. The four hours from 6:00 to 9:00 were identified as the peak commuting period. Based on space unit division and base station location, base station ID and space unit code were associated (Figure 1). By comparing the codes of space unit and residence location of a user at each hour, the departure time was determined and the base station ID matrix of origin and destination was obtained. (The decision rule is: if a user's space unit code is 0 at 6:00, and is different from the code of the residence space unit at 7:00, 6:00 is thus decided as the departure time, the code of the residence space unit is decided as the origin, and the code of space unit at 7:00 is decided as the destination. If the code of space unit at 7:00 is still 0, the departure time is further extended to the next time period till the code is not 0 and code of space unit is different from the residence code. If the space unit code at 6:00 is not 0, users with space unit codes at 6:00 and 7:00, 8:00 and 9:00, or 6:00, 7:00, 8:00, 9:00 and 11:00 are searched respectively. Whenever a change in space unit codes occurs, the different units will be decided as the origin and destination. Furthermore, an OD matrix for different hours is generated.)

Subsequently, the OD matrix of residents' travels in the case area were imported into the ABM as basic data.

Table 1. Sample of mobile phone call data.

User ID	Time of Phone Call	Base Station ID
10000001	2016-03-09-9.11.49.000000	287305843
10000002	2016-03-09-9.15.24.000000	5760859194
10000003	2016-03-09-9.15.18.000000	2872636812
10000004	2016-03-09-9.15.49.000000	2893525929
10000005	2016-03-09-9.24.13.000000	2871037354

Table 2. Sample statistics of base stations assigned to user ID at different hours of a day.

User ID	Base Station ID at 7:00	Base Station ID at 8:00	Base Station ID at 9:00	Base Station ID at 10:00	Base Station ID at 11:00
10000001	2897825643	2870117513	2870117513	2870140338	2897825643
10000002	2871865415	2871865415	2871865415	2871865415	2871865415
10000003	2870124605	2870125269	2893463025	2893410062	2893463025
10000004	2896212261	2870140337	2870129511	2870129511	2870129511
10000005	2897112172	2897112173	2897112172	2897155404	0
10000006	2896857636	2896840168	2896829356	2873044533	2896851574
10000007	2919549932	2919543433	2919549932	2919549932	2919440084

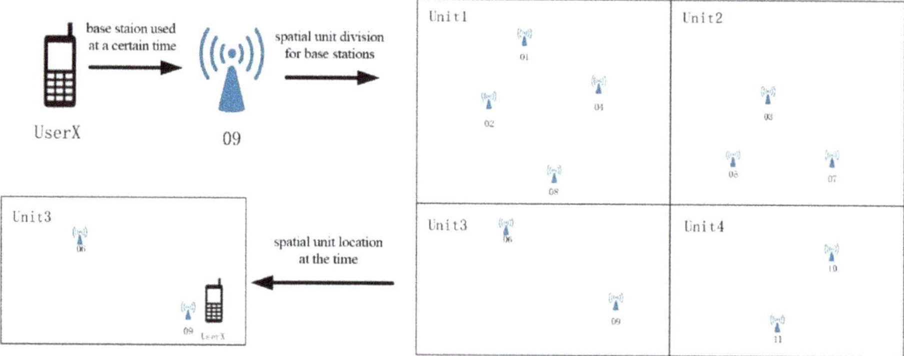

Figure 1. Procedures of assigning a user to a spatial unit.

2.2. Agent-Based Model

ABM is often used in complex giant systems such as cities. Generally speaking, a Multi-Agent System (MAS) contains many types of agents, including mobile agents such as urban residents and static agents such as urban roads. Agents run by pre-defined rules and interact with one another, producing movement and dynamic changes starting from an individual agent to the whole. As this mechanism resembles the interaction between human individuals, human and space in the city, ABM is considered as one of the best tools to understand urban functioning [43]. The model in the present study is established on the Repast S platform, and the settings of external environment and agent mobility draw reference from the open source model RepastCity [44–46]. Since residents' traveling is the only behavior studies in the research, the modeling of urban environment can be simplified into the spatial units of travel (i.e., origin and destination) as well as urban roads. Agents' behavior rules mentioned below are coded by Java and added to the RepastCity model to make it run as we designed. The rules for model running are that a resident agent moves from one spatial unit (origin) to another spatial unit (destination) at a specific time point. When the resident agent runs on the road, it may lower its speed of movement due to preset traffic congestion conditions (Figure 2).

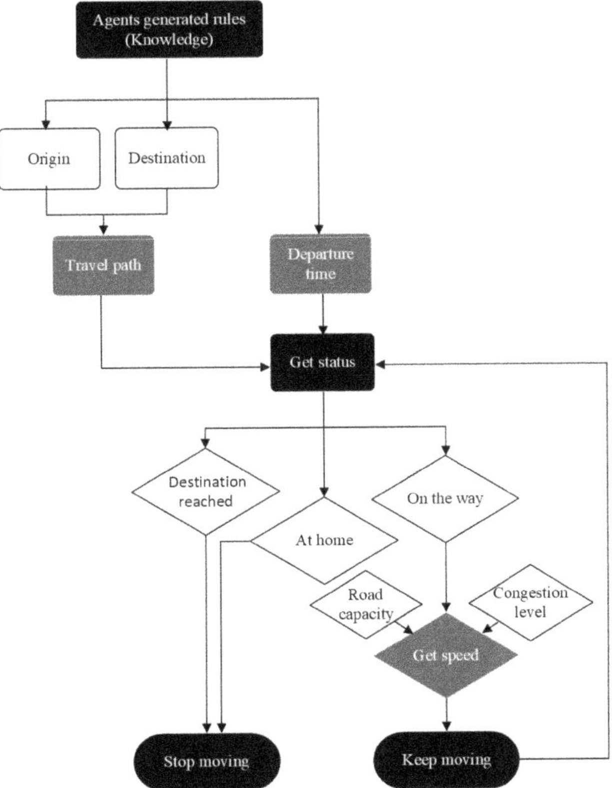

Figure 2. Diagram of rules for resident Agent behavior.

2.3. Model Hypothesis and Parameter Setting

The following hypothesis and rules are made regarding the generation of an Agent in the model and its behaviors:

First, each Agent (urban resident) generated has a specific origin and destination of travel, at a specific time point of departure. In the model, all resident Agents are generated simultaneously but are set with a specific delay value each, according to their different departure time. For example, residents depart at 6:00 have a delay value of 0 s, while those who depart at 7:00 have a delay value of 3600 s. In addition, each resident Agent is represented by a private car whose initial speed of travel is based on the driving speed of a normal motor vehicle.

Second, on each plot, a certain number of Agents is generated which is calculated using the number of residents acquired through phone data, then divided by operator's market share, and finally multiplied by the ratio of motor vehicle travel of residents.

Thirdly, traffic congestion emerges when a certain number of resident Agents concentrate in the same road intersection, and the traveling speed of residents varies in accordance to the level of congestion. Roads and nodes in the mode are generated from a shape file built in ArcGIS and are converted to Agents in RepastCity.

Fourth, residents choose the shortest route to their destination and do not change route before their arrival. The choice of path by Agents is based on the Dijkstra algorithm. Codes of space units as origin and destination are acquired in the OD matrix, and the shortest route is calculated in accordance with the road network and algorithm.

There are three major parameter variables in the commuting travel model of urban residents in the study area:

The first one is the commuter travel data of residents in each plot acquired from the OD travel matrix, and the number of travels of the corresponding Agent. The number of Agents is decided based on two factors: first, the number of residents acquired through phone data is converted to get the number of commuting residents, and is converted to get the number of travels by motor vehicles. As the Baishazhou area is located in the outskirt of the city near the Third Ring road with no subway lines, and the number of bus lines is far less than those in the inner city area, it is assumed that the majority of travels are made by private cars. The number of residents acquired through phone data is divided by the market share of the telecom operator and then the number of private car travels is acquired at a conversion factor of 1 to 1. The second factor is the ratio of the number of Agents in the simulation to the actual number of residents traveling by cars. In the statistics, we observed that there ~15,000 people traveling at the 9–10:00 period when the amount of travels is at the lowest in the study area. Previous test modeling showed that, with increasing number of Agents, the speed of simulation drops significant, while the precision of simulation results does not increase accordingly. Therefore, in the present simulation, the number of Agents is reduced so as to improve the efficiency of simulation and the traffic capacity of roads has been adjusted proportionally. The final resident-to-Agent ratio is set at 1:10, that is, one Agent represents 10 residents.

The second parameter is the speed setting. Considering the hierarchy of roads in the study area, such as urban expressways, artery roads, etc., the Agent's speed on the roads is also differentiated. In the study, two different speeds are set, i.e., the expressway speed, at 50 km/h and the artery road speed, at 30 km/h. This parameter is achieved by specifying the field of road attribute in GIS, corresponding to the speed parameters of 13.9 m/s and 8.3 m/s, respectively. The speed setting also correlates the Agent's travelling speed with the actual time unit, that is, each operation cycle (1 tick in simulation) is equal to 1 second of real time.

The third parameter is the road congestion settings. As the present study is conducted on a meso to macro scale area, roads are not categorized on a finer level, nor is the overlapping of vehicles considered. Congestion is defined by the instantaneous density of Agents on the road as vehicle density can directly demonstrate the congestion level on a road and road occupancy is often used as a quantitative indicator in traffic analysis [47]. According to the methods used in previous literature [48], the present study defines road occupancy at 0.5–1 as serious congestion, 0.3–0.5 as slight congestion, and below 0.3 as no congestion.

3. Case Study

This Baishazhou area (Location: 30.42°~30.53° N, 114.25°~114.30° E) in Wuhan was selected as the case study of commuting simulation based on the following considerations: first, Baishazhou is a new area of Wuhan, which is mostly residential in function. Therefore, the study of the impact of commuting on traffic condition in this area is of practical value. Second, since the area is still under construction and development, follow-up observations are possible to identify the differences between the simulation results of various planning programs and the traffic condition in reality. Thus, the proposed optimization and improvement measures may be of great practicality. Finally, as there is a city-level artery road in this area, i.e., the Baishazhou Avenue, the overall urban layout is distributed along the road in a belt-like shape. With frequent and extensive interactions with the surrounding areas and evident concentration of vehicle traffic, the area is deemed a valid case to evaluate the effectiveness of the model.

3.1. Data Acquisition and Processing

After the processing of mobile phone data is completed, the most important step is to allocate the residence and work places of residents at each hour to the space units. This is realized by overlapping aerial photographs, vector electronic map, and existing roads onto previously generated space units.

Working with a huge dataset like mobile phone data, the number of time division, and space unit division by different hours of a day may lead to an exponential increase in total statistical size. Therefore, except for the case study area, the division of urban space is minimized to reduce the total number of spatial units. Finally, 34 space units were generated (see Figure 3). Based on these space units, statistics were extracted for the four rush hours on each workday morning. Among these units, plots No. 1 to No. 27 were taken as the core research objects, in which commuting data of residents were acquired. Since the accurate travel routes in other plots cannot be obtained without a fine division, these plots were selected only as destinations but not origins.

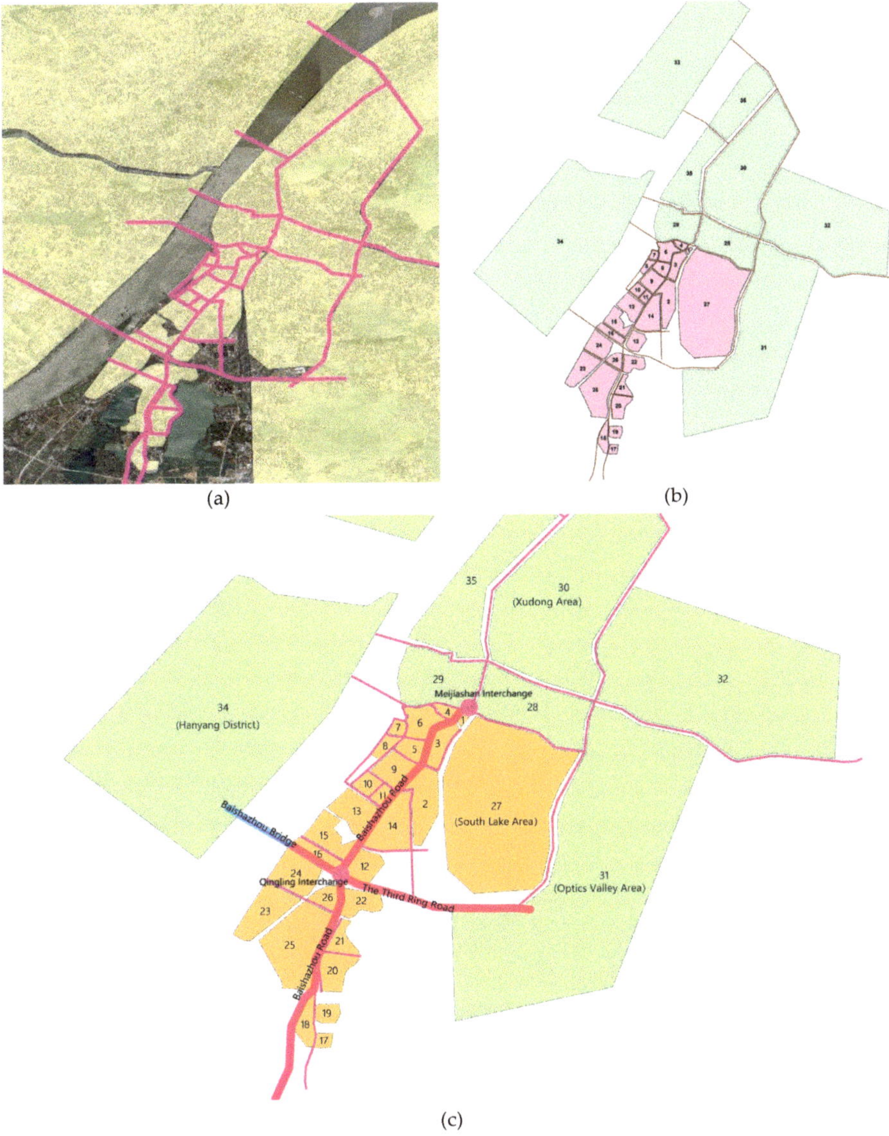

Figure 3. Road network and division of urban spatial units. (**a**) The original blocks and road network; (**b**) Simplified spatial units of land use and road network; (**c**) Names of key Roads and spatial units.

Using Monday to Friday every week as the time periods in the present study, temporal features of the residents' commute were obtained. In regular commuting, residents' travel from home to work in the morning rush hours and then from work to home during the evening rush hours. In order to reduce the amount of data in space unit division, the division of land for work is simplified. Therefore, only the commuting of residents during the morning rush hours, i.e., the four hours from 6:00–10:00, were considered in the present simulation. Using the base station data in the four hours of 6:00, 7:00, 8:00, and 9:00, the number of plots at 11:00, the number of travels at each hour and at each plot were generated and further generated an OD matrix (Table 3). The number of residents' travels in each core space unit at different hours separately was acquired (Figure 4). The ranking of the numbers for the four time slots was 7:00 > 8:00 > 6:00 > 9:00, which is consistent with our daily experience: since most employers in China set working hours between 9:00 to 17:00, residents leave home for work between 7:00 to 8:00 to reserve enough time for commuting even in face of the possible traffic jam during morning rush hours. Therefore, this period is the most popular departure time for commuters.

Table 3. Sample statistics of travel volume at each hour and at each plot.

Departure Lot No.	Destination Lot No.	Departure at 6:00	Departure at 7:00	Departure at 8:00	Departure at 9:00
1	3	20	10	10	0
1	4	20	20	20	0
1	9	10	0	0	0
1	14	0	10	0	0

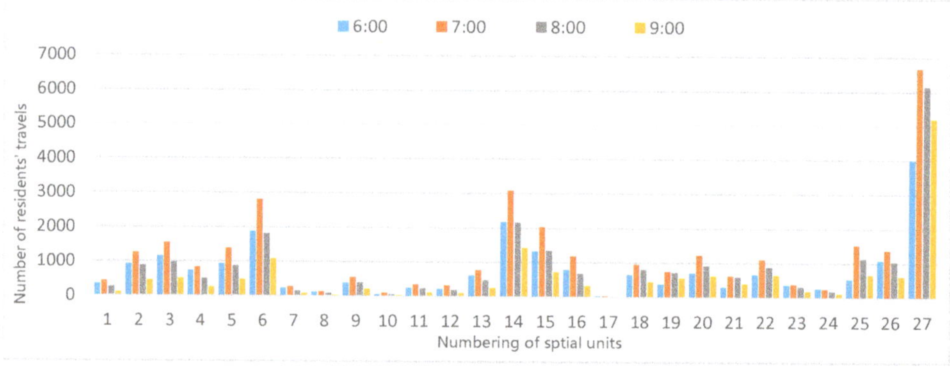

Figure 4. Comparison of residents' travels at different hours in different spatial units.

3.2. Model Simulation and Result Verification

Figure 5 presents real-time screenshot images at several time nodes during the running of the model. As the visual interface could not offer quantified traffic features, road intersections are numbered and the number of Agents at each running cycle is obtained in order to detect the occurrence, time, and level of traffic congestions. The numbering of road intersection is shown in Figure 6a.

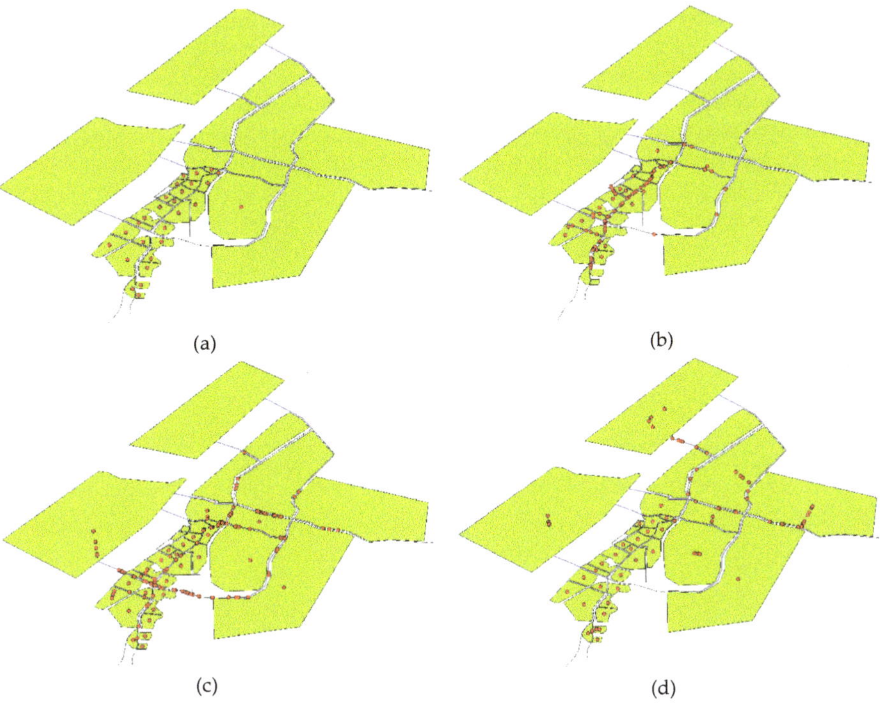

Figure 5. Real-time screenshots during model running and simulation. (**a**) Run time 30 s; (**b**) Run time 900 s; (**c**) Run time 1800 s; (**d**) Run time 3200 s.

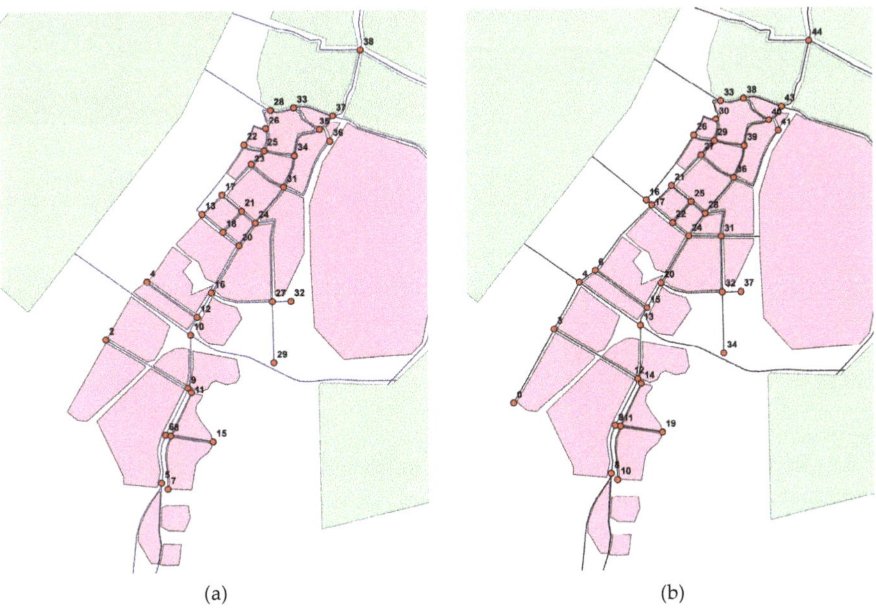

Figure 6. Numbering and distribution of road intersections. (**a**) Numbering and distribution of existing road intersections; (**b**) Numbering and distribution of planned road intersections.

Finally, the number of vehicles at each intersection in the study area was calculated in various periods each with a 90-s duration to obtain Figure 7. As shown by the changes in the number of agents at all intersections during various time periods, although differences can be seen in terms of the total number of commuter residents, the number of commuter residents on each plot, and the destinations of residents, the overall line charts generated for the four hours of study demonstrate a consistent pattern, representing similar features in the commuting of residents at each hour. Specifically, for road intersections, the peak in the line chart represents the maximum number of traffic generated, and the duration of time indicates the occurrence of congestion. Therefore, it can be seen that the most obvious congestion occurs at Intersection 10, i.e., the intersection of Baishazhou Avenue and the Third Ring Road. Other more congested intersections include No.6, No.9, and No.37.

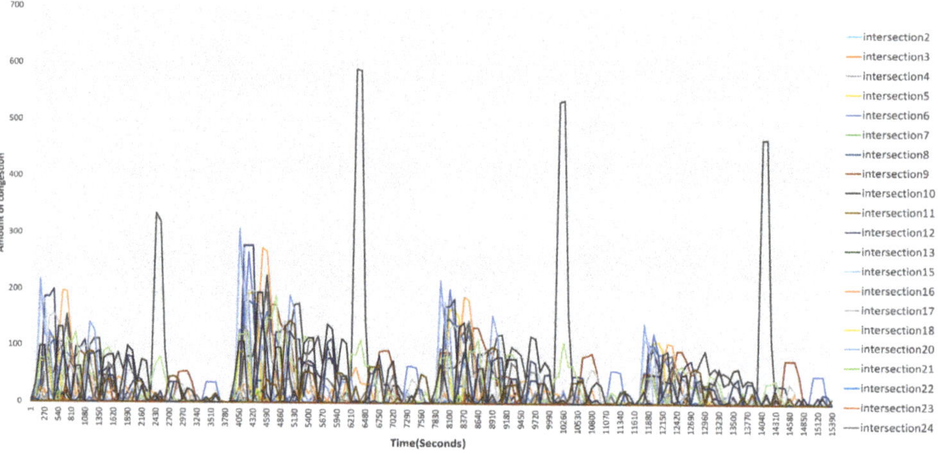

Figure 7. Changes in the number of agents at intersections.

Furthermore, the causes of congestion at each intersection can be analyzed. Combining the congestion process presentation at each intersection in the visual interface, and the number of residents departing from each block in different time periods (Table 3). Traffic in the surrounding area along the Baishazhou Avenue consists of two parts: commuting traffic from the area adjacent to Intersection 10 (the Qingling Interchange) to Plot 27 (the South Lake Area) and Plot 31 (the Optics Valley Area). While traffic in Intersection 37 (the Meijiashan interchange) comes from commuting to Plot 31 and Plot 30 (the Xudong Area). As a result, the commuting traffic flows have great impact on the road intersections near the two interchanges. In addition, as cross-river traffic in the entire area is still mainly directed to Plot 34, tension in traffic is mostly concentrated along the cross-river bridge (the Baishazhou Bridge) route.

Chinese web map providers, such as Gaode, Baidu, etc., provide not only navigation information, but also traffic forecasts based on their historical traffic data and projections of traffic conditions at different periods of a day. In the present study, traffic forecasts for the case study, the Baishazhou area, at the four time nodes, i.e., 7:00, 8:00, 9:00, and 10:00, are extracted from Gaode map as shown in Figure 8. Since these traffic forecasts are generated based on historical data, they can be considered as road traffic conditions with the highest probability of each road in the past years, and therefore, we used them in the study to verify the results of the model simulation.

Figure 8. Traffic forecasts for the four studied hours based on the big data from Gaode Map. (**a**) 7:00 traffic conditions; (**b**) 8:00 traffic conditions; (**c**) 9:00 traffic conditions; (**d**) 10:00 traffic conditions.

Since Gaode's data is only accurate to the hour and cannot be matched with the temporal granularity of traffic conditions in this simulation, comparisons can only be made on a similar time precision. Traffic at the end of each hour of simulation was used for the comparison, which means the simulation result for traffic starting at 6:00 was used for comparison with the traffic forecast at 7:00, and so on. It can be seen from the traffic forecast for Gaode that it covers only the city artery roads and above, but not the secondary roads or below, and the dark red, red, yellow, and green colors in the diagram represent increasingly better traffic conditions, from serious congestion to smooth flow. Therefore, it can be seen that the traffic condition is the worst at 8:00 when a dark red section appears at the Qingling Interchange at the intersection of Baishazhou Avenue and Third Ring Road, and the section is in yellow at the other three time nodes, indicating slight congestion. Another congestion occurs in sections along the Baishazhou Avenue ahead of and behind the Meijishan Interchange, where the heaviest traffic, in dark red color, also appears at the 8:00 time node while the sections are in yellow color at the other three time nodes. The rest of the time point is yellow. Compared to the roads in the model, intersections corresponding to the Qingling Interchange are Intersection 10, 12, and 16, while those corresponding to the Meijiahan Interchange are Intersection 35 and 37. It can be seen that most sections of congestion projected by Gaode are in accordance with the congested intersections as simulated in the model. Based on the above analysis, the results of the present model's simulation are consistent with traffic forecasts of Gaode.

4. Discussion

Statistics in the OD matrix of residents' travels at different hours show that, most of the spatial units follow the pattern of minimum traffic at 9:00, but there are also a few plots, such as Plot 27, that do not match the major pattern of travel numbers. Situated in the South Lake area and serving mainly residential functions, Plot 27 is one of the most congested areas in Wuhan during rush hours. A possible explanation is that the residents of the area choose to delay their departure time to 9:00 in order to avoid the traffic congestion period from 7:00 to 8:00, or they are stalled in traffic for too long and are considered as having not departed from the area in the statistics. This result, to some extent, verifies the significance of OD statistics and simulation by each hour: when the travel patterns of each space unit during the four hours change, traffic pressure at each intersection may also vary, and the causes behind these changes demand further analysis with the support of simulations. This is also why this study divides the unit time span of commuting travels into a one-hour basis.

In order to test the simulation results of the model for different schemes, and combined with the analysis of the causes of congestion, the roads in the study area are optimized according to the Wuhan master plan. As a measure of optimization, a waterfront north–south road along the Yangtze River and the road to the South Lake area are planned (Figure 6b). The planned and optimized road network is simulated in the model and compared with the original one. In this simulation, it is assumed that the population in this area remains unchanged and so do the places of residence and work. Comparing the simulation results of the two schemes (Figure 9), it can be seen that the optimized scheme is evidently better than the scheme before optimization: first, traffic has been distributed to multiple road intersections instead of being concentrated at an intersection before the optimization. Second, duration of traffic congestion is significantly shortened, which means congestions can be alleviated quickly even if they do occur.

According to Gaode's Traffic Report on major cities in China, 81% of them suffer from congestions during rush hours of residents' commuting [49]. Therefore, studying residents' commuting behavior as a starting point to address the wider problem of urban traffic congestion bears practical significance, not only for China but also for the world at large. The era of big data is coming. When it is less difficult to acquire data, how to use them in urban research becomes an issue that calls for deliberation [40]. Previous studies prove that mobile phone call data can more accurately reflect the commuting features of urban residents. However, most studies focus on the overall analysis of cities on a macro scale and the visual representations. Few studies are found on the micro scale dynamic analysis of residents' commuting behaviors, or on how commuting relates urban traffic.

However, as the sole data source used in the present study to understand residents' mobility, CDR data still has limitations because it is a relatively sparse data in recording the travel trajectory of residents. It is difficult to obtain the traffic mode (or speed) of residents' travel through statistical analysis. Thus, the specific correlations between commuter vehicles and mobile phone users are not discussed in the present study. Therefore, in follow-up studies, additional data sources such as bus card and traffic cameras at road intersections may facilitate the cross-examination of our research results or the setting rules of residents' commuting at a finer time-scale. Of course, these rely on the availability of data, which remain difficult to collect compared with other sources at present.

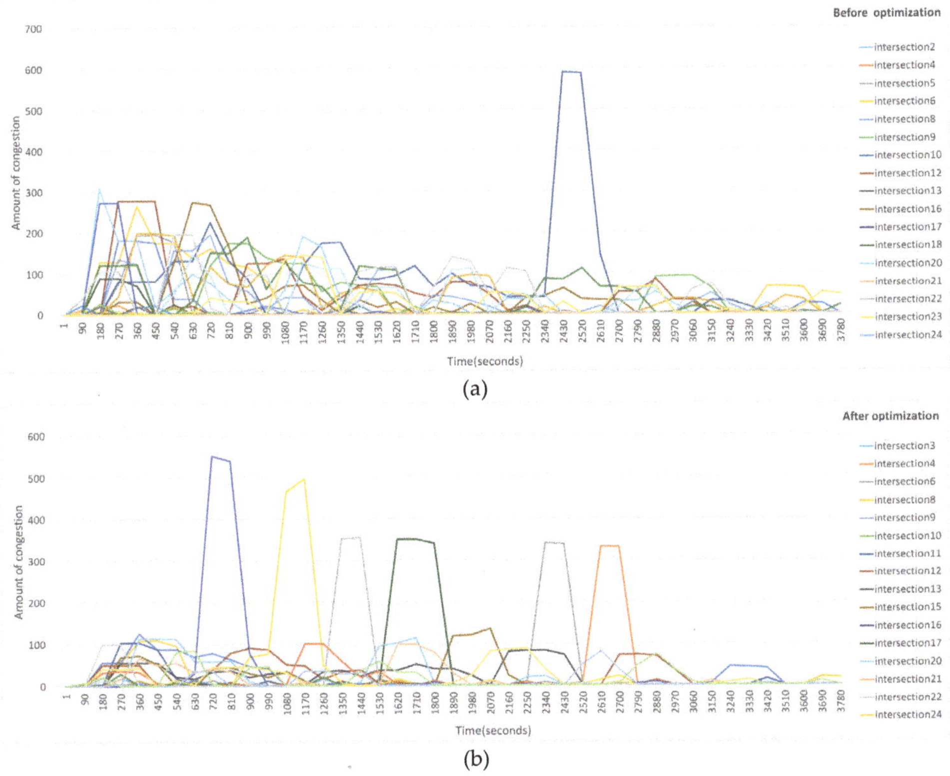

Figure 9. Comparison of traffic conditions before and after the optimization of the road network. (**a**) The traffic conditions of road intersections before optimization; (**b**) The traffic conditions of road intersections after optimization.

5. Conclusions

In the present study, the Agent-based model is used to simulate traffic condition during commuting hours in a local urban area. First, the commuting demand of residents calculated by mobile phone data is used to simulate congestions on the existing urban road network. Then, data backtracking is used to identify the causes of congestion and to analyze the simulation results. Finally, the results of simulation are proven to be consistent with the actual traffic conditions. Although the data used are simplified for the easiness of processing and modeling, the study is still believed to be a positive endeavor of combining big data and ABM in an urban study, and it offers a valuable approach to studying residents' commuting and urban traffic.

The approach used in the paper has several limitations that merit future consideration: first, vehicles other than commuter cars, such as buses are not considered. Prospective studies are expected to incorporate other available data sources and machine learning approaches to further specify modes of commuter travels and incorporate buses as a major means of transportation. Second, in the present model's construction, lanes and traffic flow directions on the roads are not specified. In a congestion setting, only the density of vehicles in a certain section is considered while the overlapping of vehicles is neglected. This means that the model cannot sufficiently reflect traffic conditions in reality and also leads to the fact that the simulation results cannot be analyzed on a finer scale for deduction of the processes. Prospective studies are expected to further refine the road and traffic systems of the model.

Author Contributions: Conceptualization, Y.Y.; Methodology, L.L.; Validation, H.W. and L.L.; Formal Analysis, H.J.; Data Curation, Z.P.; Writing—Original Draft Preparation, H.W.; Writing—Review & Editing, Y.Y.; Project Administration, H.W.; Supervision, Z.P.; Funding Acquisition, H.W. and Q.N.

Funding: The research was funded by the China Postdoctoral Science Foundation (No. 2016M600609), the MOE Layout Foundation of Humanities and Social Sciences (No. 19YJCZH187); the National Natural Science Fund for Young Scholars (No. 51708425); the Natural Science Fund of China (No. 51778503); and the National Natural Science Fund for Young Scholars (No. 51708426).

Acknowledgments: The authors acknowledge the contribution of all the anonymous reviewers that improved the quality of the paper.

Conflicts of Interest: The authors declare no conflict of interest.

References

1. Vickrey, W.S. Congestion Theory and Transport Investment. *Am. Econ. Rev.* **1969**, *59*, 251–260.
2. Zhou, J.; Murphy, E.; Long, Y. Commuting efficiency in the Beijing metropolitan area: An exploration combining smartcard and travel survey data. *J. Transp. Geogr.* **2014**, *41*, 175–183. [CrossRef]
3. Scott, D.M.; Kanaroglou, P.S.; Anderson, W.P. Impacts of commuting efficiency on congestion and emissions: Case of the Hamilton CMA, Canada. *Transp. Res. Part D Transp. Environ.* **1997**, *2*, 245–257. [CrossRef]
4. Friedman, M.S.; Powell, K.E.; Hutwagner, L.; Graham, L.M.; Teague, W.G. Impact of changes in transportation and commuting behaviors during the 1996 Summer Olympic Games in Atlanta on air quality and childhood asthma. *Jama J. Am. Med. Assoc.* **2001**, *285*, 897–905. [CrossRef] [PubMed]
5. Brueckner, J.K. Urban Sprawl: Diagnosis and Remedies. *Int. Reg. Sci. Rev.* **2000**, *23*, 160–171. [CrossRef]
6. Arnott, R.; De Palma, A.; Lindsey, R. A structural model of peak-period congestion—A traffic bottleneck with elastic demand. *Am. Econ. Rev.* **1993**, *83*, 161–179.
7. Anas, A.; Xu, R. Congestion, land use, and job dispersion: A general equilibrium model. *J. Urban Econ.* **1999**, *45*, 451–473. [CrossRef]
8. Gonzales, E.J.; Daganzo, C.F. Morning commute with competing modes and distributed demand: User equilibrium, system optimum, and pricing. *Transp. Res. Part B Methodol.* **2012**, *46*, 1519–1534. [CrossRef]
9. Chatman, D.G. Does TOD Need the T? *On the Importance of Factors Other Than Rail Access. J. Am. Plan. Assoc.* **2013**, *79*, 17–31. [CrossRef]
10. Salon, D. Neighborhoods, cars, and commuting in New York City: A discrete choice approach. *Transp. Res. Part A Policy Pract.* **2009**, *43*, 180–196. [CrossRef]
11. Fosgerau, M.; Kim, J.; Ranjan, A. Vickrey meets Alonso: Commute scheduling and congestion in a monocentric city. *J. Urban Econ.* **2018**, *105*, 40–53. [CrossRef]
12. Song, C.M.; Koren, T.; Wang, P.; Barabasi, A.L. Modelling the scaling properties of human mobility. *Nat. Phys.* **2010**, *6*, 818–823. [CrossRef]
13. Ratti, C.; Frenchman, D.; Pulselli, R.M.; Williams, S. Mobile Landscapes: Using Location Data from Cell Phones for Urban Analysis. *Environ. Plan. B Plan. Des.* **2006**, *33*, 727–748. [CrossRef]
14. Wesolowski, A.; Eagle, N.; Tatem, A.J.; Smith, D.L.; Noor, A.M.; Snow, R.W.; Buckee, C.O. Quantifying the Impact of Human Mobility on Malaria. *Science* **2012**, *338*, 267–270. [CrossRef] [PubMed]

15. Bengtsson, L.; Lu, X.; Thorson, A.; Garfield, R.; Von Schreeb, J. Improved Response to Disasters and Outbreaks by Tracking Population Movements with Mobile Phone Network Data: A Post-Earthquake Geospatial Study in Haiti. *PLoS Med.* **2011**, *8*, 8. [CrossRef] [PubMed]

16. Calabrese, F.; Colonna, M.; Lovisolo, P.; Parata, D.; Ratti, C. Real-Time Urban Monitoring Using Cell Phones: A Case Study in Rome. *Ieee Trans. Intell. Transp. Syst.* **2011**, *12*, 141–151. [CrossRef]

17. Aasa, A. Application of mobile phone location data in mapping of commuting patterns and functional regionalization: A pilot study of Estonia. *J. Maps* **2013**, *9*, 10–15.

18. Kung, K.S.; Greco, K.; Sobolevsky, S.; Ratti, C. Exploring Universal Patterns in Human Home-Work Commuting from Mobile Phone Data. *PLoS ONE* **2014**, *9*, e96180. [CrossRef] [PubMed]

19. Tu, W.; Cao, J.; Yue, Y.; Shaw, S.L.; Zhou, M.; Wang, Z.; Chang, X.; Xu, Y.; Li, Q. Coupling mobile phone and social media data: A new approach to understanding urban functions and diurnal patterns. *Int. J. Geogr. Inf. Sci.* **2017**, *31*, 2331–2358. [CrossRef]

20. Yue, Y.; Zhuang, Y.; Yeh, A.G.; Xie, J.Y.; Ma, C.L.; Li, Q.Q. Measurements of POI-based mixed use and their relationships with neighbourhood vibrancy. *Int. J. Geogr. Inf. Syst.* **2017**, *31*, 658–675. [CrossRef]

21. Doyle, J.; Hung, P.; Farrell, R.; McLoone, S. Population Mobility Dynamics Estimated from Mobile Telephony Data. *J. Urban Technol.* **2014**, *21*, 109–132. [CrossRef]

22. Pei, T.; Sobolevsky, S.; Ratti, C.; Shaw, S.L.; Li, T.; Zhou, C. A new insight into land use classification based on aggregated mobile phone data. *Int. J. Geogr. Inf. Sci.* **2014**, *28*, 1988–2007. [CrossRef]

23. Pei, T.; Sobolevsky, S.; Ratti, C.; Shaw, S.L.; Li, T.; Zhou, C. Development of origin-destination matrices using mobile phone call data. *Transp. Res. Part C Emerg. Technol.* **2014**, *40*, 63–74.

24. Calabrese, G.F.; Lorenzo, D.; Liu, L.; Ratti, C. Estimating origin-destination flows using mobile phone location data. *IEEE Pervasive Comput.* **2011**, *10*, 264–323. [CrossRef]

25. Alexander, L.; Jiang, S.; Murga, M.; González, M.C. Origin–destination trips by purpose and time of day inferred from mobile phone data. *Transp. Res. Part C Emerg. Technol.* **2015**, *58*, 240–250. [CrossRef]

26. Yue, Y.; Lan, T.; Yeh, A.G.; Li, Q.Q. Zooming into individuals to understand the collective: A review of trajectory-based travel behaviour studies. *Travel Behav. Soc.* **2014**, *1*, 69–78. [CrossRef]

27. Çolak, S.; Lima, A.; González, M.C. Understanding congested travel in urban areas. *Nat. Commun.* **2016**, *7*, 10793. [CrossRef]

28. Yao, Y.; Hong, Y.; Wu, D.; Zhang, Y.; Guan, Q. Estimating the effects of "community opening" policy on alleviating traffic congestion in large Chinese cities by integrating ant colony optimization and complex network analyses. *Comput. Environ. Urban Syst.* **2018**, *70*, 163–174. [CrossRef]

29. Nguyen, Q.T.; Bouju, A.; Estraillier, P. Multi-agent Architecture with Space-time Components for the Simulation of Urban Transportation Systems. *Procedia Soc. Behav. Sci.* **2012**, *54*, 365–374. [CrossRef]

30. Doniec, A.; Mandiau, R.; Piechowiak, S.; Espié, S. A behavioral multi-agent model for road traffic simulation. *Eng. Appl. Artif. Intell.* **2008**, *21*, 1443–1454. [CrossRef]

31. Ciari, F.; Balmer, M.; Axhausen, K.W. A new mode choice model for a multi-agent transport simulation. In Proceedings of the 8th Swiss Transport Research Conference, Ascona, Switzerland, 6–8 October 2008.

32. Knapen, L.; Keren, D.; Cho, S.; Bellemans, T.; Janssens, D.; Wets, G. Analysis of the Co-routing Problem in Agent-based Carpooling Simulation. In *Ant 2012 And Mobiwis 2012*; Shakshuki, E., Younas, M., Eds.; Elsevier Science Bv: Amsterdam, The Netherlands, 2012; pp. 821–826.

33. Kaddoura, I.; Kickhöfer, B.; Neumann, A.; Tirachini, A. Optimal Public Transport Pricing: Towards an Agent-based Marginal Social Cost Approach. *J. Transp. Econ. Policy* **2016**, *49*, 200–218.

34. Dimitrov, S.; Ceder, A.; Chowdhury, S.; Monot, M. Modeling the interaction between buses, passengers and cars on a bus route using a multi-agent system. *Transp. Plan. Technol.* **2017**, *40*, 592–610. [CrossRef]

35. Liu, J.; Kockelman, K.M.; Boesch, P.M.; Ciari, F. Tracking a system of shared autonomous vehicles across the Austin, Texas network using agent-based simulation. *Transportation* **2017**, *44*, 1261–1278. [CrossRef]

36. Lu, M.; Hsu, S.C. Spatial Agent-based model for environmental assessment of passenger transportation. *J. Urban Plan. Dev.* **2017**, *143*, 04017016. [CrossRef]

37. Bellemans, T.; Bothe, S.; Cho, S.; Giannotti, F.; Janssens, D.; Knapen, L.; Körner, C.; May, M.; Nanni, M.; Pedreschi, D.; et al. An Agent-Based Model to Evaluate Carpooling at Large Manufacturing Plants. In *Ant 2012 And Mobiwis 2012*; Shakshuki, E., Younas, M., Eds.; Elsevier Science Bv: Amsterdam, The Netherlands, 2012; pp. 1221–1227.

38. Calabrese, F.; Ferrari, L.; Blondel, V.D. Urban Sensing Using Mobile Phone Network Data: A Survey of Research. *Acm Comput. Surv.* **2014**, *47*, 25. [CrossRef]
39. Batty, M.; Axhausen, K.W.; Giannotti, F.; Pozdnoukhov, A.; Bazzani, A.; Wachowicz, M.; Ouzounis, G.; Portugali, Y. Smart cities of the future. *Eur. Phys. J. Spec. Top.* **2012**, *214*, 481–518. [CrossRef]
40. Wu, H.; Liu, L.; Yu, Y.; Peng, Z. Evaluation and Planning of Urban Green Space Distribution Based on Mobile Phone Data and Two-Step Floating Catchment Area Method. *Sustainability* **2018**, *10*, 214. [CrossRef]
41. Song, C.; Qu, Z.; Blumm, N.; Barabási, A.L. Limits of Predictability in Human Mobility. *Science* **2010**, *327*, 1018–1021. [CrossRef]
42. Yang, C.; Zhang, Y.; Ukkusuri, S.V.; Zhu, R. Mobility Pattern Identification Based on Mobile Phone Data. In *Transportation Analytics in the Era of Big Data*; Ukkusuri, S.V., Yang, C., Eds.; Springer International Publishing: Basel, Switzerland, 2012; pp. 217–232.
43. Levy, S.; Martens, K.; Van Der Heijden, R. Agent-based models and self-organisation: : Addressing common criticisms and the role of agent-based modelling in urban planning. *Town Plan. Rev.* **2016**, *87*, 321–338. [CrossRef]
44. Malleson, N. Extending RepastCity. 2012. Available online: https://code.google.com/p/repastcity/wiki/ExtendingRepastCity3 (accessed on 23 July 2019).
45. Malleson, N. RepastCity-Model Structure. 2012. Available online: https://code.google.com/p/repastcity/wiki/RC3ModelStructure (accessed on 23 July 2019).
46. Malleson, N. RepastCity-A Demo Virtual City. 2012. Available online: https://code.google.com/p/repastcity/wiki/RepastCity3 (accessed on 23 July 2019).
47. Kopf, J.; Ishimaru, J.M.; Nee, J.; Hallenbeck, M.E. Central Puget Sound Freeway Network Usage and Performance, 2003 Update. Highway Traffic Control. 2005. Available online: http://www.highwaytraffic.com.au/#!/ (accessed on 23 July 2019).
48. Widyantoro, D.H.; Munajat, M.E. Fuzzy traffic congestion model based on speed and density of vehicle. In Proceedings of the 2014 International Conference of Advanced Informatics: Concept, Theory and Application, Bandung, Indonesia, 20–21 August 2014.
49. Gaode Map. Traffic Report on Major Cities in China 2017. 2018. Available online: http://cn-hangzhou.oss-pub.aliyun-inc.com/download-report/download/yearly_report/2017/ (accessed on 23 July 2019).

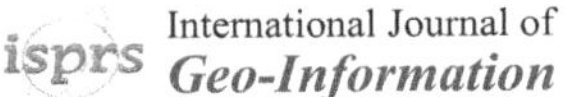
International Journal of
Geo-Information

Article

Heuristic Bike Optimization Algorithm to Improve Usage Efficiency of the Station-Free Bike Sharing System in Shenzhen, China

Zhihui Gu [1,2], Yong Zhu [1], Yan Zhang [1,*], Wanyu Zhou [1] and Yu Chen [1]

1   College of Architecture and Urban Planning, Shenzhen University, Shenzhen 518060, China;
    gzh@szu.edu.cn (Z.G.); 2172322425@email.szu.edu.cn (Y.Z.); 2161140409@email.szu.edu.cn (W.Z.);
    szuchenyu@szu.edu.cn (Y.C.)
2   Shenzhen Key Laboratory for Optimizing Design of Built Environment, Shenzhen 518060, China
*   Correspondence: zhyan@szu.edu.cn; Tel.: +86-755-26732869

Received: 8 April 2019; Accepted: 17 May 2019; Published: 21 May 2019

Abstract: Station-free bike sharing systems (BSSs) are a new type of public bike system that has been widely deployed in China since 2017. However, rapid growth has vastly outpaced the immediate demand and overwhelmed many cities around the world. This paper proposes a heuristic bike optimization algorithm (HBOA) to determine the optimal supply and distribution of bikes considering the effect of bicycle cycling. In this approach, the different bike trips with separate bikes can be connected in space and time and converted into a continuous trip chain for a single bike. To improve this cycling efficiency, it is important to properly design the bicycle distribution. Taking Shenzhen as an example, we implement the algorithm with OD matrix data from Mobike and Ofo, the two large bike sharing companies which account for 80% of the shared bike market in Shenzhen, over two days. The HBOA results are as follows. 1) Only one-fifth of the bike supply is needed to meet the current usage demand if the bikes are used efficiently, which means a large number of shared bikes in Shenzhen remain in an idle state for long periods. 2) Although the cycling demand is high in many areas, it does not mean that large numbers of bikes are needed because the continuous inflow caused by the cycling effect of bikes will meet most of the demand by itself. 3) The areas with the highest demands for optimal bikes are residential, followed by industrial, public transportation, official and commercial areas, on both working and non-working days. This algorithm can be an objective basis for city related departments to manage station-free BSSs and be applied to design the layout of bikes in small-scale spatial units to help station-free BSSs operate efficiently and minimize the need to relocate the bikes without reducing the level of user satisfaction.

Keywords: station-free BSS; HBOA; oversupply; use efficiency

---

1. Introduction

The station-free bike sharing system (BSS), also known as the free-floating or fourth generation BSS, is a new type of public bike system that has been widely deployed in China since 2017 and expanded to other countries. In this system, bikes can be selected using private apps and parked in the appropriate places. Comparing the traditional station-based BBS, the station-free BSS can expand the bike sharing service with lower cost as the high initial capital investment required for the docking stations is not needed. Due to the freedom and convenience the BSS provides, it has attracted a large number of consumers requiring "last mile" transportation.

However, rapid growth has vastly outpaced the immediate demand and overwhelmed Chinese cities, where infrastructures and regulations were not prepared to handle a sudden flood of millions of shared bikes [1]. In many cities, adequate parking facilities for bikes are not available, city officials lack the regulation experience for this mode of transportation, and normal social behaviors have not been established [2]. It is very common to have more than one operating company for the station-free BSS in a city. From an operational perspective, the most important goal is to occupy the market, which is why many companies would prefer to provide more bikes and exceed the demand [3]. On the other hand, large bike fleets are associated with a waste of resources because many bikes would remain idle for long periods, making the system inefficient.

In response to these problems, more and more Chinese cities such as Shanghai, Hangzhou, Guangzhou, Shenzhen, have banned the addition of further shared bikes [4]. A series of strict regulations for bike share providers are being implemented in China and European countries, including orderly parking, quality and timely maintenance of the bikes, license system for operators and fleet size control [5,6]. However, a fundamental unsolved problem is determining how many station-free bikes are sufficient to effectively meet the needs of users?

The Bike Sharing Planning Guide provides guidelines for the fleet size of a station-based BSS, which are 10–16 stations per km^2, 10–30 bicycles for every 1000 residents within coverage area, and 202.5 docking spaces for every bicycle [7]. However, they are for the station-based BSS, not the station free BBS. Moreover, these quantities are rough indications and mainly depend on the characteristics of city built-environment, such as land use, population density, and road conditions.

To determine the optimal fleet size and distribution of station-free BBS, this paper proposes a heuristic bike optimization algorithm (HBOA) considering the efficiency of bike cycling. It can be an objective basis for city related departments to issue the total control policy and be applied to design the layout of bikes in small-scale spatial units to improve the system's efficiency.

2. Literature Review

In past decades, many studies have focused on three main issues related to public bike systems with docking stations: the spatial structure of a city [8–11], the inflow and outflow of vehicles at each station [12–14], and the rebalancing of the vehicles among stations [15–18]. In a station-based BSS, the supply of the vehicles must be compatible with the scale of the fixed stations. Once the construction of the stations is complete, the system is difficult to change. Therefore, extensive research on station-based BSSs has focused on the locations and capacities of stations to optimize the efficiency of these systems [19–22].

Station-free BSSs completely differ from station-based BSSs. The characteristics of a station-free BSS allow the system scale to be enlarged by providing many vehicles without station capacity constraints. Because vehicle parking is scattered and the spatial distribution is changing all the time, the demand for rebalancing might increase in some cases, and predictions of potential imbalances are relatively complex. There are no predetermined stations in these systems, so scheduling schemes are often unclear, even if the real-time parking distribution is known. Furthermore, sometimes relocation occurs based on spur-of-the-moment changes without following a specific strategy [23].

Most research on relocation in station-free BSSs has extended the ideas and methods applied to stationed-based BSSs, and studies have focused on the effects of urban features [24,25], spatiotemporal patterns of biking behavior [26,27], and relocation or rebalancing of shared bikes [28–31]. For example, by setting virtual traffic zones, each traffic area is treated as a bike sharing station, and the first distribution and relocation scheme of the BSS are designed according to the demand model combined with the vehicle outflows and inflows in the traffic zone. Some studies have proposed algorithms to achieve efficient relocation strategies for stationed-based BSSs from both static and dynamic perspectives [32,33]. Other studies used OD matrix data from bike sharing companies to analyze and simulate bike sharing travel patterns [34]. In another study, the demand was forecasted with deep learning methods to predict the gap between the inflow and outflow of sharing bike trips at a TAZ [35]. These studies based on virtual stations have helped simplify the analysis process, but they fail to take full advantage of the unique use characteristics of free-floating BSSs to a large extent. First, due to the randomness of parking with no docking stations, it is difficult to set a fixed TAZ for relocation.

In the division of virtual traffic zones, zones that are too large may not reflect the reality of operation, and zones that are too small will make relocation complicated. Second, a very important difference between a stationed-based BSS and station-free BSS is that the chain of travel can more easily occur at a smaller scale because of the spontaneous usage in the station-free BSS.

Due to the large number and usage frequency of shared bikes, the randomness of shared bikes movement and spacing is high. From the perspective of complex systems, the behaviors of users can be regarded as a self-organizing process. On the one hand, the hidden reasons behind user behaviors are worth studying compared to the inherent system randomness. On the other hand, it is important to identify which factors in the complex system are critical to the self-organizing process. For example, Chen et al. simulated the interactions between supply and demand based on agent-based modeling and suggested that the key aspects of the sustainable development of the bicycle-sharing market are twofold: the reliability of the supply must be improved, and the uncertainty in the demand must be reduced. Standardizing the distribution of shared bikes and fixing their locations could solve the disorder issue associated with excessive supply [36]. Vazifeh et al. proposed a solution to address the minimum fleet-size problem at the urban scale for the general case of taxi trips based on the demand mobility [37]. This study combined applied mathematics and graph algorithms from computer science field and transformed the minimum fleet problem into a minimum path coverage problem based on the directed graphs, which led to breakthroughs in operational efficiency. If the chain of travel is considered, it is possible to optimize and simplify the relocation of bikes and improve the efficiency of the station-free BSS. However, unlike taxis, the principle of shared bikes is that individuals can use bikes "as-needed" by finding the surrounding bikes instead of dispatching vehicles on demand. Taxi drivers can actively choose the optimal route, but a shared bike must be selected by a user according to the location and parking time and is controlled by the user.

Therefore, based on the construction of a shared bike trip chain with actual riding data for a certain period of time, this paper develops a heuristic algorithm to determine the optimal demand for public bikes with little operation intervention required. This method is then applied for multi-company cycling data analysis in the megacity of Shenzhen, China. The results indicate that the algorithm can reveal the mobility patterns of shared bikes and provide useful information for shared bikes to improve the use efficiency at the city scale.

3. Methodology

Similar to the solution for the minimum fleet-size problem, the purpose of this study is to improve the operational efficiency of a shared bike system by constructing a shared bike trip chain. In areas with high cycling requirements, it is not always necessary to supply more bikes. If the number of cycling-in bikes is always greater than the number of cycling-out bikes, then it means that the demand does not exceed the supply. The more bikes there are in a system, the greater the inefficiency of the shared bikes. As shown in Figure 1, there are six consecutive cycling trips among the three sites. In the ideal scenario, one bike at site A is sufficient for all trips. However, in the oversupply scenario, for example, two bikes are required at each site, and the six trips may be completed by up to six different bikes. However, no matter how many different bikes are being used, the bike stock at site A is always greater than 1, and the numbers of bikes at sites B and C are always greater than 2. When the volume of shared bikes is greater than the cycling requirement, bikes will remain unused, and road space will be wasted. Within a certain time interval and space range, the number of bikes in stock is always greater than zero, regardless of the possibility of damage to the bikes; therefore, the supply is greater than the demand, and there are no more bikes potentially needed. The key to improving the self-organization process of cycling is to fix the initial positions of the shared bikes at the optimal positions.

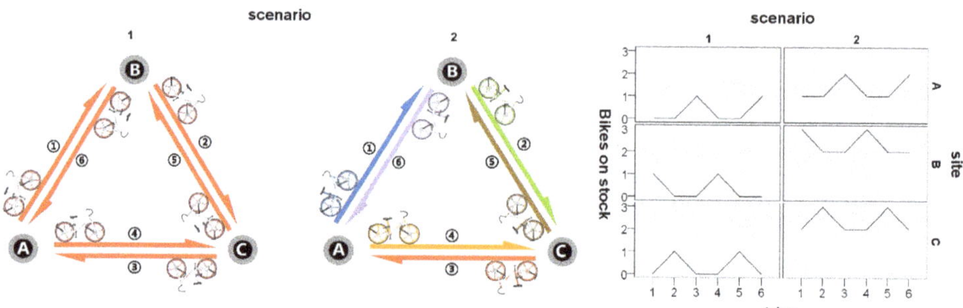

Figure 1. Bike movement and stocks in different scenarios: (1) only one bike at site A and (2) two bikes at each site.

Fixed boundaries are not suitable for shared bikes because of the random nature of user behavior and the unrestricted parking of station-free bikes. Therefore, we propose a heuristic bike optimization algorithm (HBOA). The core concept of the HBOA is to use the fewest number of bikes to meet all cycling requirements. The principle of using shared bikes is "first come, first served". If the ending position of one trip is close to the starting position of another trip, the ending time of the last trip and the starting time of the next trip can be continuous in time; thus, in theory, the same bike can be used for both trips.

To obtain a more reasonable number of optimized bikes, we set the minimum time interval for cycling requirements between the ending time of the last trip and the starting time of next trip to 10 min, and the maximum Euclidean distance between the ending position of the last trip and the starting position of the next trip is 100 m. That is, after completing the last trip, the optimized bike would service the closest trip at that time within 100 m of the ending position. Finally, the number of optimized bikes could be considered the ideal delivery scale of shared bikes in meeting all cycling requirements. The initial positions of these bikes can also be considered an optimal configuration for delivering or dispatching the shared bikes.

The calculation process of the HBOA is shown in Figure 2. We set all the data from valid cycling trips as data set C, including O, D, Ts, and Te information. O is the original position of trip Ci, D is the destination position of trip Ci, Ts is the starting time of trip Ci, and Te is the ending time of trip Ci. First, one of the earliest cycling trips is selected randomly and recorded as the first trip for optimized bike Bj,m(O, D, Ts, Te), where (j = 1, m = 1). Then, the trips within 100 m of Bj,m(D) are searched, and the closest trip at given starting time is identified as the next trip Bj, m + 1. This process continues until it is impossible to identify another trip for this optimized bike. The search for the earliest cycling trips in the unmarked cycling data set continues. The first trip for a new optimized bike is identified as Bj,m(O, D, Ts, Te), where (j = j + 1, m = 1). All subsequent trips are also analyzed. The process of searching is repeated until each trip is marked as one trip for an optimized bike. Obviously, the result of this algorithm is not unique. However, considering the size of the data set and the aim of the HOBA, the result does not need to be the best solution to improve the usage efficiency of shared bikes. The time-space distribution characteristics of these optimized bikes can be used as a configuration reference for initial bike delivery.

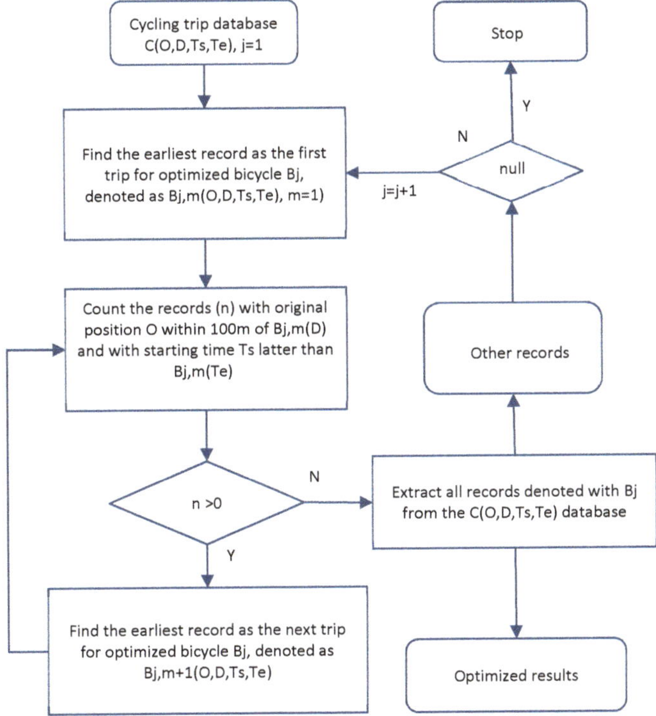

Figure 2. The calculation process of the HBOA.

4. Study Area and Data Preprocessing

Shenzhen, the youngest megacity in China, was founded only 40 years ago. By the end of 2017, the city had 12.52 million people in an area of 1997.27 km^2 [38]. According to a report, there were approximately 10 shared bike companies in Shenzhen with approximately 890 thousand shared bikes in the market in August 2017. In September, a new shared bike policy was released by the Shenzhen government that suspended the launch of new shared bike systems in the city [39].

Through the API ports of shared bike apps, the positions of all vacant bikes are given in real time. Therefore, we scanned the positions of vacant bikes for two companies, Ofo and Mobike, which account for more than 80% of the shared bike market. Limited by the app client, we only obtained 2 days of scanning data from 6–7 May 2018. These dates fall on a Sunday and Monday, representing non-working and working days. The weather conditions were similar on these two days, with sporadic light rain. We found approximately 306 thousand different Mobike bikes and 434 thousand Ofo bikes by scanning the entire city, accounting for over 80% of the total number of shared bikes.

Because it took approximately ten minutes to scan the entire city, the time interval of scanning was ten minutes. By comparing the positions of the vacant bikes at different times, it can be determined whether a bike moved, and the origin-destination positions and trip times can be obtained. Correspondingly, we can obtain the Euclidean distance and speed of these trips. However, there may be two types of data errors. The first type of error is equipment error. According to an actual test, the error of the GPS for a vacant bike returned to the same position can reach approximately 100 m. The second type of error is inference error. For example, some shared bike companies use motor vehicles to manually dispatch bikes, and the speed of cycling trips is too fast in these cases. Additionally, it is also possible that some bikes are missed during the scanning process, resulting in a long trip time. Therefore, data cleaning was performed for the original data. First, trips with Euclidean distances less than 200 m were considered invalid, or walking was considered a more reasonable alternative. Second, trips with an average speed greater than 25 km/h may involve the manual dispatching of bikes by motor vehicles instead of normal cycling. Other trips with low speeds are indistinguishable and were retained for use in the HBOA. After cleaning, only 640 thousand available movements remained, and the average usage time of each bike was less than one. Nearly 340 thousand shared bikes did not move in two days.

In addition, two types of databases were used in this paper, as shown in Figure 3. One database includes the transportation routes in Shenzhen 2018, as well as the metro stations and bus stations. The other database includes building information from 2015, such as outline and usage information for residential buildings, urban village buildings, industrial buildings, commercial buildings, official buildings, and others. Among these buildings, urban village buildings are a special type of low-cost residential building in Shenzhen. These data will help us further analyze the temporal and spatial distribution characteristics of optimized bike use. Urban area in Shenzhen has gradually transformed from a belt shape within the original Special Economic Zone (including Luohu, Futian, Nanshan and Yantian districts) into an outward radial shaped city in the past three decades, which, to some extent, deviates a multi-center development pattern [40]. Six central areas are selected to compare with the spatial distribution of shared bikes. Three of them are public service centers, including Baoan center area, Futian center area and Luohu center area. Two commercial centers are Nanshan center area and Huaqiang center area. One is an official employment center, High-tech center area in Nanshan district.

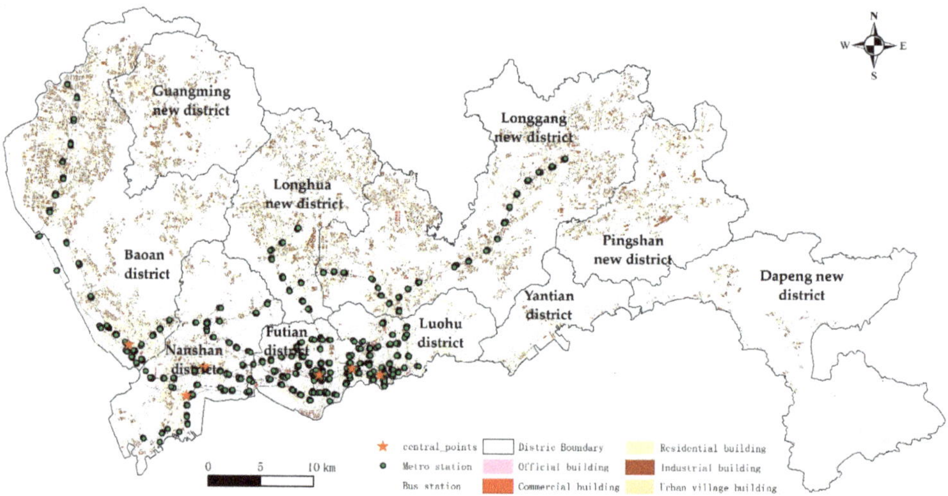

Figure 3. The transportation and building information in Shenzhen.

5. Results and Discussion

5.1. Optimized and Actual Bike Availability

The HBAO indicated that only 137,216 bikes were needed to complete all valid trips on 6 May 2018, and 154,625 bikes were needed on 7 May 2018. The average usage number of an optimized bike on each day was 4.6 and 4.2. Overall, less than 1/5 of all shared bikes were used.

As shown in Figure 4, there are bikes in almost every land unit (200 m * 200 m) in Shenzhen built environment. However, over 99% of these units, the actual number of bikes is higher than the number of optimized bikes, which indicates that the supply is higher than the demand. In particular, the number of bikes in the central area exceeds the number of optimized bikes by more than 100.

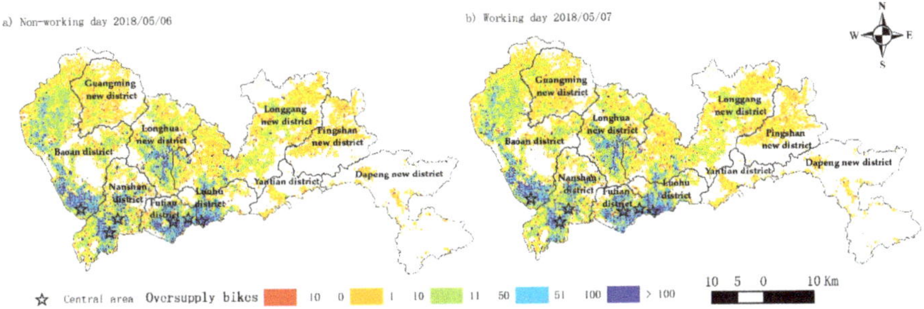

Figure 4. Difference between the actual number of available bikes and the number of optimized bikes in two days.

We took all the exits of the Houhai metro station as an example, Houhai metro station is located the central area of Nanshan district of Shenzhen, which is surrounded by commercial and residential buildings, and close to some popular public spaces, such as Shenzhen Bay Stadium, Shenzhen Bay Park and Shenzhen Talent Park. We counted the shared bikes within 100 m of the metro station exits which cycling in or out every 10 min on 6 May 2018. There were 883 cycling-in and 997 cycling-out bikes. The initial number of shared bikes around this station was 735 at 0:00 a.m., and there were always more than 500 bikes available in 24 h. As shown in Figure 5, this station had a serious oversupply issue.

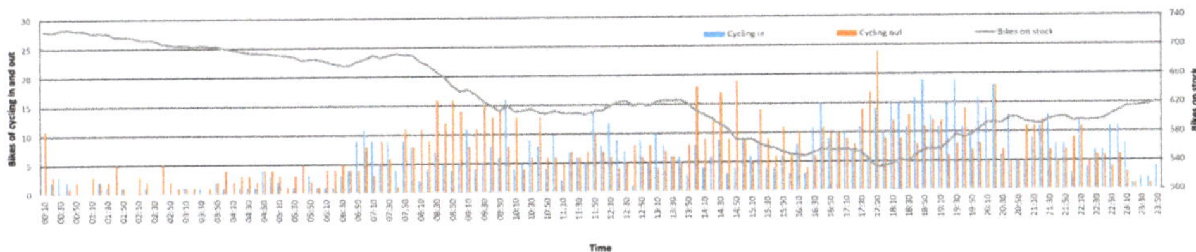

Figure 5. Cycling-in and cycling-out bikes and stock changes around the Houhai metro station every 10 min.

The result of the HOBA showed that only 219 optimized bikes are needed around Houhai station exits. The idling of a large number of bikes is a waste of resources and road space. High cycling requirements do not necessarily correspond to the need for more bikes, especially in areas where the cycling requirements re self-balanced by user activities. Shared bicycle companies tend to delivery more bikes in high cycling requirement area to occupy the market. However, for areas with a higher frequency of use, if the cycling in and out could reach equilibrium, more delivery means less efficiency. It is more worthwhile to see where the bikes heading to these areas come from. As mentioned earlier, the key to improving the self-organization process of cycling is to fix the initial positions of the shared bikes at the optimal positions. Therefore, we would compare the high requirements space of cycling and the spatial distribution of optimized bikes' initial positions in the next section.

5.2. Spatial Requirements of Cycling and Spatial Distribution of Optimized Bikes' Initial Positions

We used the kernel density estimation to compare the requirement space and the ideal supply space of shared bikes. We defined the origin positions distribution of all valid trips as the requirement space of cycling, and the initial positions distribution of all optimized bikes as the ideal space of supply demand. As shown in Figure 6, we find that the requirement spaces are similar on working days and non-working days, and the correlation coefficient was 0.942 ($p < 0.001$). The supplying demand spaces optimized bikes on working days and non-working days has a high correlation coefficient too (0.862, $p < 0.001$). We picked the areas with expected values of greater than 25 uses per hectare as high requirement areas and those with expected values of greater than 5 bikes per hectare as the high supply areas for optimized bikes. It can be seen that these areas are consistent or adjacent to the central areas of each district in Shenzhen.

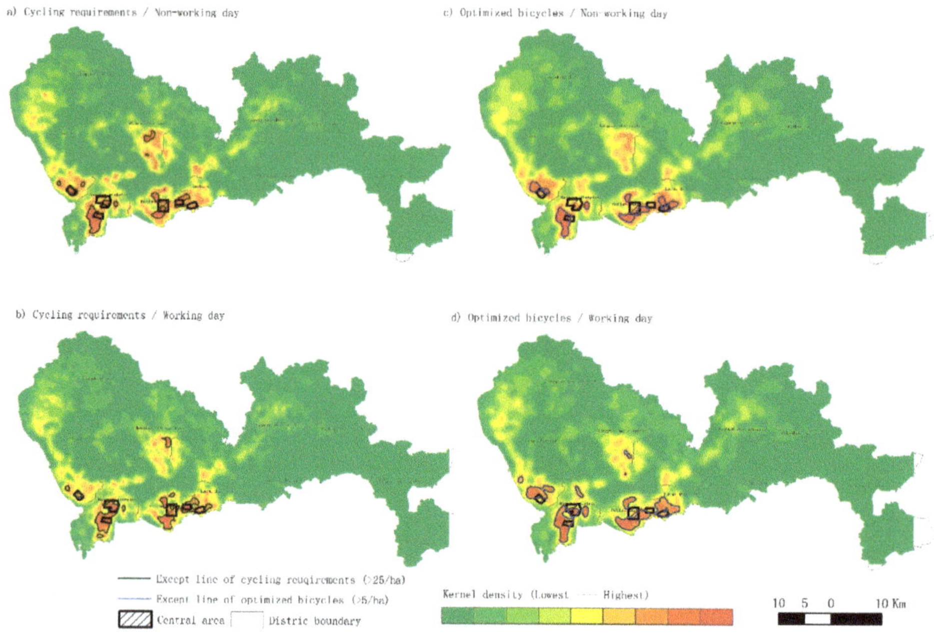

Figure 6. Spatial requirements of cycling and the spatial distribution of optimized bikes.

Overlay analysis is applied for these spaces, including overlays of the transportation and building data (Figure 7). The overlay results suggest that: (1) the area with high requirement for cycling is more consistent with the central areas of the city. Except for the central area of Luohu, other high requirement areas basically contain the central areas; (2) most of areas with high requirement for cycling are not necessarily consistent with high supply-demand space, but adjacent, such as Baoan and Futian central areas; (3) There are also some very stable areas with high demand and supply both in working or non-working days, especially in Nanshan district. It is easy to understand that the central areas often bring a lot of cycling requirements because of its high vitality. And due to its non-residential properties and attractive features to the surrounding area, a large number of cycling in bikes could meet the cycling requirements without the need of a large supply of shared bikes. One of the distinguishing features of the Nanshan District, which is different from other central areas, is that the number of metro stations and lines through it are less than those of other districts. But it is still difficult to explain why some areas have higher stability of supply demand than others. And these areas should be our most noteworthy space, because the initial bikes in these areas would result in higher efficiency. In the next section, we will focus on the initial position of each optimized bike and its surrounding traffic and the built environment.

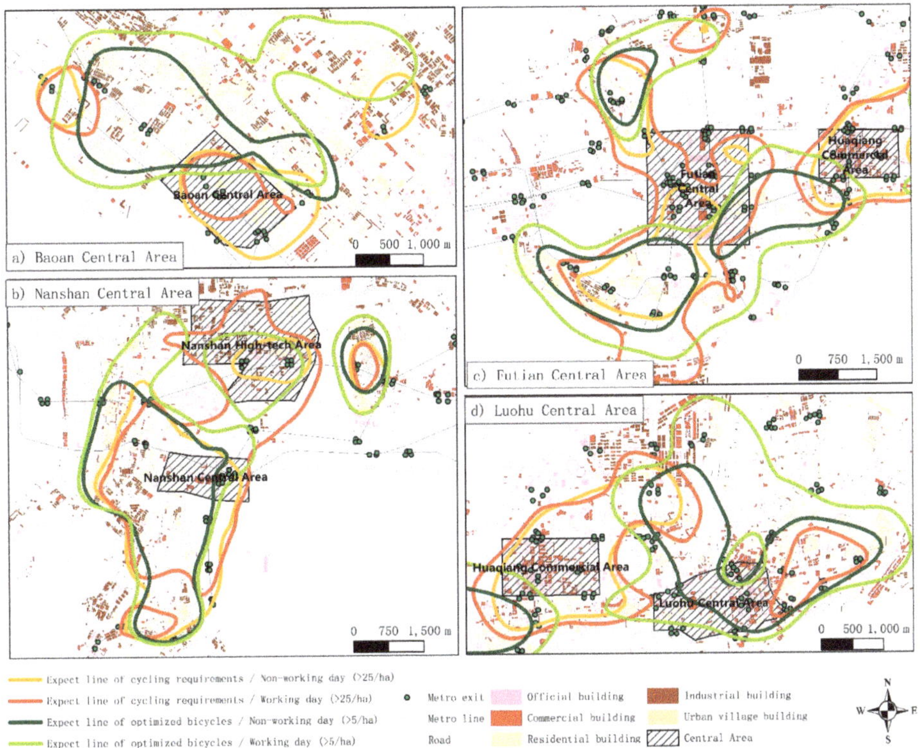

Figure 7. Built environments in six central areas with high requirement or supply spaces for shared bikes.

5.3. The Temporal and Spatial Characteristics of the Initial Position of Each Optimized Bike

In this section, the temporal and spatial characteristics of the initial positions of all optimized bikes are discussed. There are two main reasons for assigning an optimized bike: the departure time of cycling out is relatively early, or the numbers of cycling in bikes couldn't meet the demand for cycling out. Therefore, finding the initial departure time and its surrounding built environment of these optimized bikes could help us better understand their supply needs. In order to easy statistics, we set a simple proximity priority for optimized bikes. First of all, the optimized bikes closest to the public transport facility are considered as demand of transfer. Among the remaining optimized bikes, public transportation is preferred too. Metro connections are assumed for those bikes within 100 m of all metro station exits. Bus connections are assumed for those within 50 m of all bus stations. Finally, the closest building to each remaining unused bike is assumed to be related to the use of that bike.

As shown in Table 1, about 45% of optimized bikes are closest to residential buildings and urban village buildings. This is because most of the first trip in one day starts from the residence. What's interesting is that the area nearby industrial buildings also a significant need for optimized bikes. Although the metro stations have higher cycling requirements as mentioned by other literature [41], only 5% of optimized bikes is needed within 100 m of all metro stations. The previous analysis in Section 5.1 also proved it.

Table 1. The number and percentage of optimized bikes nearby public transportation facilities and different buildings.

Near Type	05/06		05/07	
	Optimized Bikes	**Percentage**	**Optimized Bikes**	**Percentage**
Near metro stations (<100 m)	6902	5.13%	7478	4.91%
Near bus stations (<50 m)	10,693	7.95%	12,664	8.32%
Closest to residential building	34,519	25.65%	33,266	21.86%
Closest to urban village building	29,293	21.77%	35,141	23.09%
Closest to industrial building	24,660	18.32%	25,575	16.80%
Closest to official building	8893	6.61%	11,589	7.61%
Closest to commercial building	7784	5.78%	9894	6.50%
Closest to other building	11,838	8.80%	16,590	10.90%
Total	134,582	100%	152,197	100%

Combining the nearby spatial characteristics and temporal characteristics of the first trip for all optimized bikes, we obtained Figure 8. In addition to the early peak at 7–9 a.m., there is also a small peak during the night from 0:00 to 1:00 a.m. This peak is partially because the algorithm searches for the earliest trip starting at 0:00 a.m., another reason may be the public transportation stoppage and high taxi prices during the nighttime. Another finding is that industrial buildings, like living buildings, have the same night peaks and early peak demand both on working and non-working day. One possible explanation is that these factories implement a three-shift switching working system which resulted in higher demand for optimized bikes at midnight and early peak time. In general, the distribution of optimized bikes is mainly in areas where the first trip of cycling out earlier or the number of cycling in bikes is less than the demand for cycling out. Correspondingly, major destinations for cycling in, such as commercial buildings and official buildings have less demand for optimized bikes.

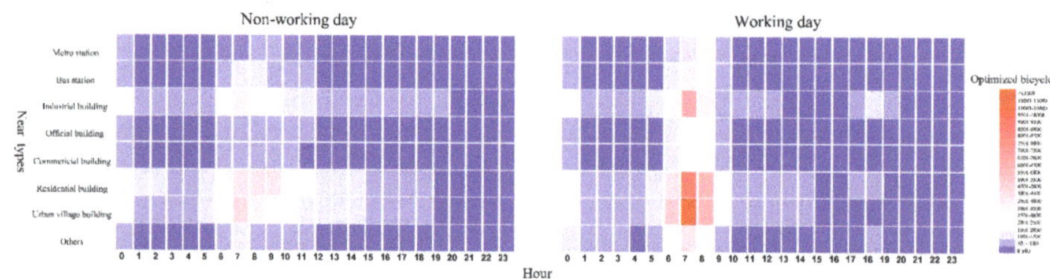

Figure 8. The nearby spatial and temporal characteristics of the first trip for all optimized bikes.

Furthermore, we compare the spatial distributions of optimized bikes in various nearby areas to identify the specific characteristics of the spatial demand for optimized bikes.

As shown in Figure 9, on working and non-working days, the spatial distribution of optimized bikes near public transportation facilities displays some spatial characteristics. The metro stations around the central areas have relatively high optimized bike demands on both working and non-working days, especially in Nanshan district. Our study found that 53.3% of the employed population in Nanshan high-tech area lives within 5 km. However, the layout of metro lines in Nanshan district is seriously mismatched with the commuter corridor [42]. The bus line has similar problems, mainly along the east-west strip, while the commuter corridor in Nanshan district is north-south. The high demand for optimized bikes at these public transportation facilities shows that the direct accessibility of public transportation is poor and require more transfer in the last mile.

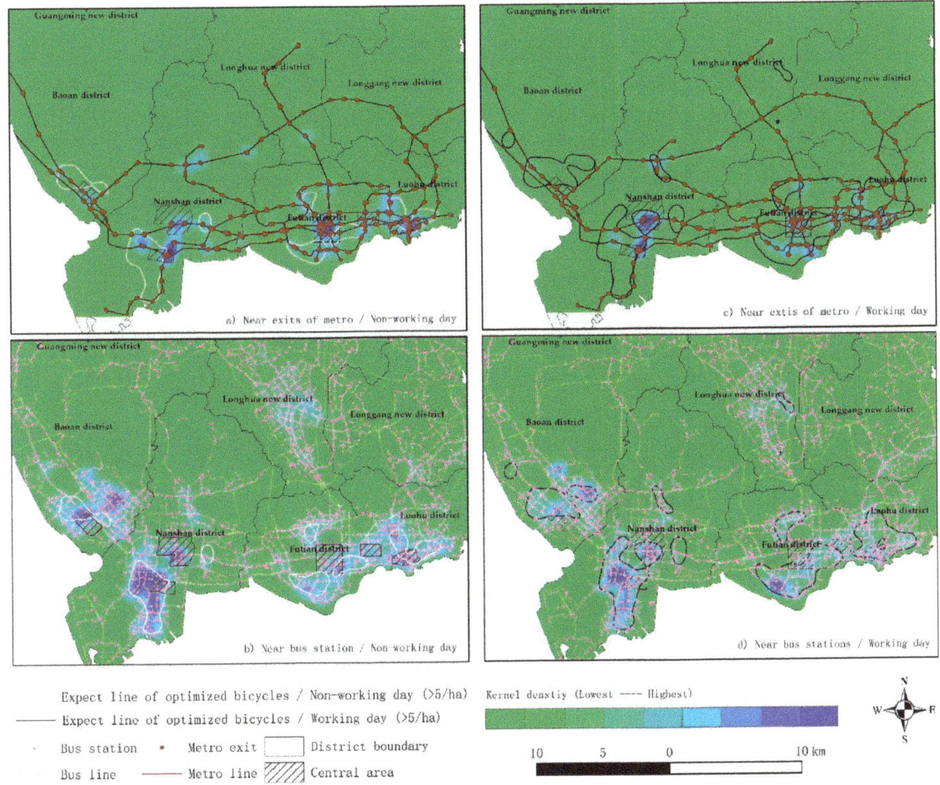

Figure 9. The spatial distribution of optimized bikes near public transportation facilities.

Similar to the previous analysis, we compared the spatial distribution of optimized bikes in adjacent buildings to find these relatively stable areas with high demand for shared bikes. As shown in Figure 10a), urban village buildings next to the central areas have a significantly high demand for optimized bikes. There is no such obvious spatial characteristic in residential buildings (Figure 10b), except for the buildings in Nanshan District. Among the industrial buildings, Bantian industrial zone in Longhua District is very special area which is an industrial production base for electronic information, biotechnology and new materials in Shenzhen (Figure 10c). Whether the three-shift working system generally occurs here needs further investigation. For official and commercial buildings, there are also some such particularly stable areas with relatively high demand for optimized bikes both on weekdays and non-working days (Figure 10d,e).

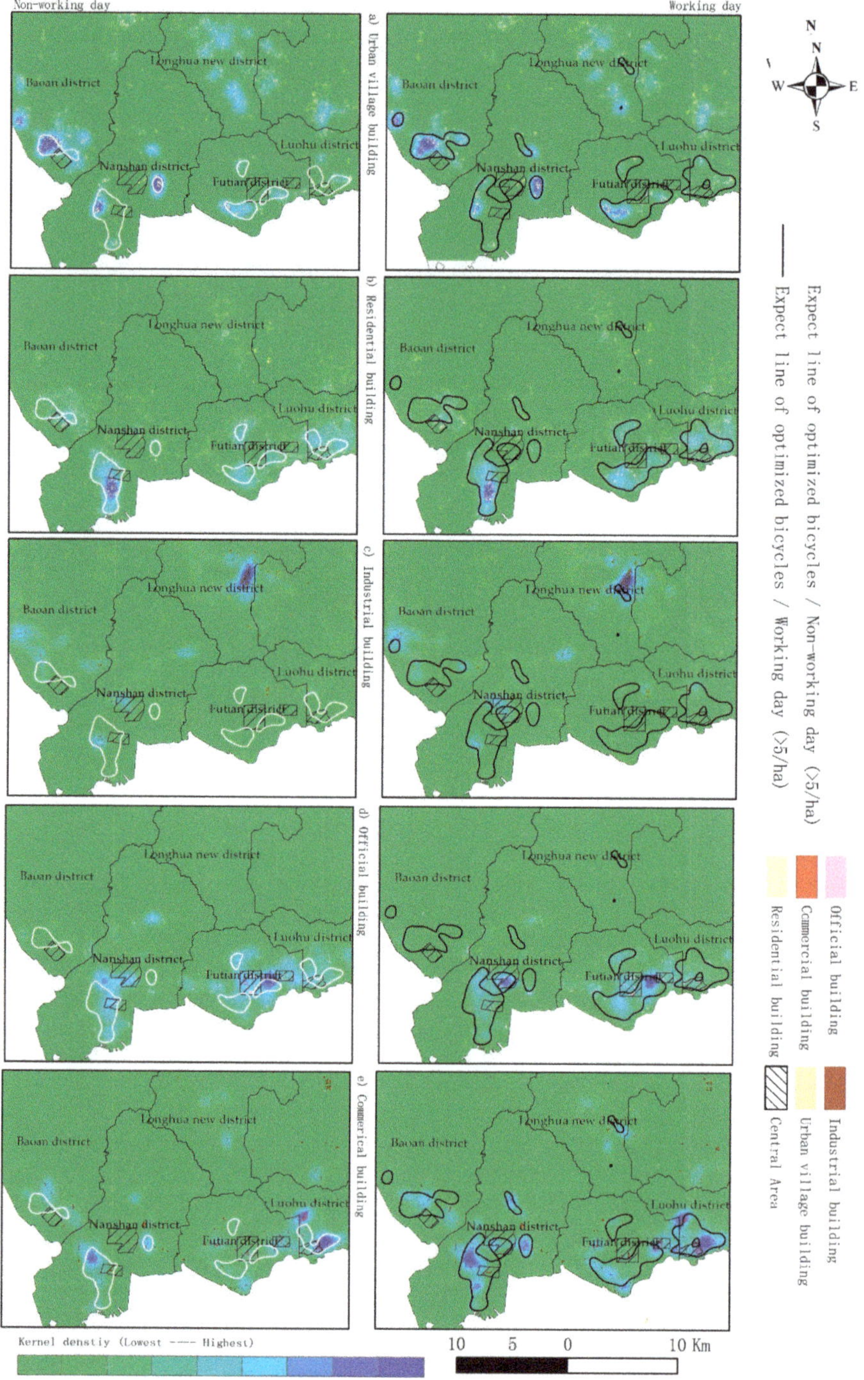

Figure 10. Spatial distribution of optimized bikes near different buildings.

Due to data limitations, we only analyzed the spatial distribution of optimized bikes on two days and found that the results exhibited high consistency on both working and non-working days. If the proposed algorithm was applied to a long-term data, more spatial characteristics may be identified to help us understand the complementary relationship between public transportation and shared bikes or direct shared bikes for more scientific and effective delivery.

6. Conclusions

The boom of station-free BSSs has increased customer convenience but also caused inefficiency due to the excessive supply of bikes. It presents regulation challenges for city officials. What is the optimal fleet size of the station-free BSS which can fully meet the needs of users and make bicycles be used efficiency as well? How should these bikes be spatially distributed on this supply scale? This paper, which is based on actual travel data from the station-free BSS in Shenzhen, proposes an algorithm to construct a travel chain and determine the optimal bike demands in different spatial units.

Our results show that in Shenzhen city, only one-fifth of shared bikes is needed to meet the current usage demand if the bikes are used efficiency. With a population of 12.52 million in Shenzhen in 2017, the average number of bikes per 1000 people is 13 vehicles, which is in the range of 10–30 vehicles/1000 people recommended by the Bike Sharing Planning Guide. Correspondingly, our optimized results increase the average usage number of each bikes from below 1 to above 4, which greatly improved the efficiency of shared bikes.

Our results also identify some areas with the high spatial requirements of cycling and the ideal spatial distribution of optimized bikes' initial positions. If the initial distribution is established according to this demand, the trips that occur throughout the day can be completed with as few bikes as possible without reducing the level of user satisfaction. Obviously, the spatial distribution of bikes will change dramatically at the end of the day. In response to this situation, the operator can relocate the bikes to the initial distribution using a static strategy at night. Thus, this approach establishes both a benchmark for the layout of station-free bikes and a target strategy for relocation.

The proposed HBOA is simple in principle, and the calculations are convenient to perform. Although the calculation results may not be optimal at all times, this information can be used to significantly improve the use efficiency of shared bikes. Thus, the results could be used by companies to meet the maximum coverage demand with the smallest number of bikes and as a tool for urban planners to scientifically manage the station-free BSS. From the perspective of the city as a whole, the total supply of shared bicycles should be kept at an optimal level to improve the overall operational efficiency of the urban traffic system. In this sense, it is necessary to break the barriers between different operators of the overall station-free BSS and enable users to rent and return bikes among different station-free BSSs. The two-day analysis results reflected the stability of bike use patterns and some specific differences between working and non-working days. If long-term data from more companies could be analyzed, the results would be more reliable and further improve the system efficiency by minimizing the size of the shared bike with the HBOA. In this case, additional physical infrastructure is not needed, but the current infrastructure could be more intelligently managed.

Author Contributions: Methodology, Zhihui Gu; Validation, Yu Chen; Data Curation, Yong Zhu and Wanyu Zhou; Writing—Review & Editing, Zhihui Gu and Yan Zhang.

Funding: This research was funded by the National Natural Science Foundation of China, grant No. 51778366.

Acknowledgments: The authors would like to thank the anonymous reviewers and academic editors for their helpful comments on an earlier draft of this paper.

Conflicts of Interest: The authors declare no conflict of interest. The funders had no role in the design of the study; in the collection, analyses, or interpretation of data; in the writing of the manuscript, or in the decision to publish the results.

References

1. Taylor, A. The Bike-Share Oversupply in China: Huge Piles of Abandoned and Broken Bicycles. The Atlantic. Available online: https://www.theatlantic.com/photo/2018/03/bike-share-oversupply-in-china-huge-piles-of-abandoned-and-broken-bicycles/556268/ (accessed on 28 November 2018).
2. Sisson, P. Free Ride: Is Bike Share's Next Evolution a System without Stations? Curbed. 31 March 2017. Available online: https://www.curbed.com/2017/3/31/15141002/cycling-transportation-bike-share-bluegogo-ride (accessed on 19 July 2017).
3. Li, X.H. Ten Ministries and Commissions Established New Rules and Regulations, Can the Barbaric Growth of Sharing Bicycles Be Controlled? Available online: http://news.sohu.com (accessed on 10 August 2017).
4. Xinmin Evening News. From Barbaric Growth to Total Control, Shared Bike Chaos Improved Significantly. 11 November 2019. Available online: https://news.sina.com.cn/o/2019-01-11/doc-ihqhqcis5270576.shtml (accessed on 1 May 2019).
5. Institute for Transportation & Development Policy. *The Bike Sharing Planning Guide*; Institute for Transportation and Development Policy: New York, NY, USA, 2013.
6. European Bicycle Manufactures Association. Bike Sharing in China. Available online: http://ebma-brussels.eu/bike-sharing-in-china/ (accessed on 29 September 2017).
7. European Bicycle Manufactures Association. Bike Sharing in Europe. Available online: http://ebma-brussels.eu/bike-sharing-in-europe/ (accessed on 29 September 2017).
8. Froehlich, J.; Neumann, J.; Oliver, N. Measuring the pulse of the city through shared bicycle programs. In Proceedings of the International Workshop on Urban, Community, and Social Applications of Networked Sensing Systems, UrbanSense08, Raleigh, NC, USA, 4 November 2008.
9. Vogel, P.; Greiser, T.; Mattfeld, D.C. Understanding bike-sharing systems using data mining: Exploring activity patterns. *Procedia Soc. Behav. Sci.* **2011**, *20*, 514–523. [CrossRef]
10. Lee, K.H.; Ko, E.J. Relationships between neighbourhood environments and residents' bicycle mode choice: A case study of Seoul. *Int. J. Urban Sci.* **2014**, *18*, 383–395. [CrossRef]
11. Cui, Y.; Mishra, S.; Welch, T.F. Land use effects on bicycle ridership: A framework for state planning agencies. *J. Transp. Geogr.* **2014**, *41*, 220–228. [CrossRef]
12. Saharidis, G.K.D.; Fragkogios, A.; Zygouri, E. A multi-periodic optimization modeling approach for the establishment of a bike sharing network: A case study of the City of Athens. In Proceedings of the International MultiConference of Engineers and Computer Scientists, Hong Kong, China, 12–14 March 2014; Volume 2.
13. Chen, Q.; Sun, T. A model for the layout of bike stations in public bike-sharing systems. *J. Adv. Transp.* **2015**, *49*, 884–900. [CrossRef]
14. Corcoran, J.; Li, T.; Rohde, D.; Charles-Edwards, E.; Mateo-Babiano, D. Spatio-temporal patterns of a Public Bicycle Sharing Program: The effect of weather and calendar events. *J. Transp. Geogr.* **2014**, *41*, 292–305. [CrossRef]
15. Sayarshad, H.; Tavassoli, S.; Zhao, F. A multi-periodic optimization formulation for bike planning and bike utilization. *Appl. Math. Model.* **2012**, *36*, 4944–4951. [CrossRef]
16. Martinez, L.; Caetano, L.; Eiró, T.; Cruz, F. An optimisation algorithm to establish the location of stations of a mixed fleet biking system: An application to the city of Lisbon. *Procedia Soc. Behav. Sci.* **2012**, *54*, 513–524. [CrossRef]
17. Frade, I.; Ribeiro, A. Bike-sharing stations: A maximal covering location approach. *Transp. Res. Part A Policy Pract.* **2015**, *82*, 216–227. [CrossRef]
18. Caggiani, L.; Camporeale, R.; Marinelli, M.; Ottomanelli, M. User satisfaction based model for resource allocation in bike-sharing systems. *Transp. Policy* **2017**, 1–10. [CrossRef]
19. Krykewycz, G.R.; Puchalsky, C.M.; Rocks, J.; Bonnette, B.; Jaskiewicz, F. Defining a primary market and estimating demand for major bicycle-sharing program in Philadelphia, Pennsylvania. *Transp. Res. Rec. J. Transp. Res. Board* **2010**, *2143*, 117–124. [CrossRef]
20. Lin, J.; Yang, T. Strategic design of public bicycle sharing systems with service level constraints. *Transp. Res. Part E Logist. Transp. Rev.* **2011**, *47*, 284–294. [CrossRef]
21. García-Palomares, J.C.; Gutiérrez, J.; Latorre, M. Optimizing the location of stations in bike-sharing programs: A GIS approach. *Appl. Geogr.* **2012**, *35*, 235–246. [CrossRef]

22. Lin, J.; Yang, T.; Chang, Y. A hub location inventory model for bicycle sharing system design: Formulation and solution. *Comput. Ind. Eng.* **2013**, *65*, 77–86. [CrossRef]
23. Rei, S. Demand Modeling and Relocation Strategies for Free-Floating Bike Sharing Systems. 2017. Available online: https://athene-forschung.unibw.de/doc/121232/121232.pdf (accessed on 20 September 2017).
24. Zhao, J.; Deng, W.; Song, Y. Ridership and effectiveness of bike sharing: The effects of urban features and system characteristics on daily use and turnover rate of public bikes in China. *Transp. Policy* **2014**, *35*, 253–264. [CrossRef]
25. El-Assi, W.; Mahmoud, M.; Habib, K. Effects of built environment and weather on bike sharing demand: A station level analysis of commercial bike sharing in Toronto. *Transportation* **2017**, *44*, 589–613. [CrossRef]
26. Zhou, X. Understanding spatiotemporal patterns of biking behavior by analyzing massive bike sharing data in Chicago. *PLoS ONE* **2015**, *10*, e0137922. [CrossRef] [PubMed]
27. Faghih-Imani, A.; Eluru, N. Incorporating the impact of spatio-temporal interactions on bicycle sharing system demand: A case study of new york citibike system. *J. Transp. Geogr.* **2016**, *54*, 218–227. [CrossRef]
28. Caggiani, L.; Camporeale, R.; Ottomanelli, M. A dynamic clustering method for relocation process in free-floating vehicle sharing systems Leonardo. *Transp. Res. Procedia* **2017**, *27*, 278–285. [CrossRef]
29. Bulhões, T.; Subramanian, A.; Erdoğan, G.; Laporte, G. The static bike relocation problem with multiple vehicles and visits. *Eur. J. Oper. Res.* **2018**, *264*, 508–523. [CrossRef]
30. Legros, B. Dynamic repositioning strategy in a bike-sharing system; how to prioritize and how to rebalance a bike station. *Eur. J. Oper. Res.* **2019**, *272*, 740–753. [CrossRef]
31. Liu, J.; Sun, L.; Chen, W.; Xiong, H. Rebalancing Bike Sharing Systems: A Multi-source Data Smart Optimization. In Proceedings of the 22nd ACM SIGKDD International Conference on Knowledge Discovery and Data Mining, San Francisco, CA, USA, 13–17 August 2016; ACM: New York, NY, USA; pp. 1005–1014. [CrossRef]
32. Pal, A.; Zhang, Y. Free-floating bike sharing: Solving real-life large-scale static rebalancing problems. *Transp. Res. Part C* **2017**, *80*, 92–116. [CrossRef]
33. Caggiani, L.; Camporeale, R.; Ottomanelli, M.; Szeto, W.Y. A modeling framework for the dynamic management of free-floating bike-sharing systems. *Transp. Res. C Emerg. Technol.* **2018**, *87*, 159–182. [CrossRef]
34. Reiss, S.; Bogenberger, K. GPS-Data Analysis of Munich's Free-Floating Bike Sharing System and Application of an Operator-based Relocation Strategy. In Proceedings of the 2015 IEEE 18th International Conference on Intelligent Transportation Systems (ITSC), Las Palmas, Spain, 15–18 September 2015; pp. 584–589.
35. Xu, C.; Ji, J.; Liu, P. The station-free shared bike demand forecasting with a deep learning approach and large-scale datasets. *Transp. Res. Part C* **2018**, *95*, 47–60. [CrossRef]
36. Chen, X.; Qu, Q.; Chen, M.-H.; Fang, S.; Cheng, Y. The sustainable existence of China's bicycle-sharing market: to oversupply or to disappear. *Sustanability* **2018**, *10*, 4214. [CrossRef]
37. Vazifeh, M.M.; Santi, P.; Resta, G.; Strogatz, S.H.; Ratti, C. Addressing the minimum fleet problem in on-demand urban mobility. *Nature* **2018**, *557*, 534–538. [CrossRef] [PubMed]
38. Shenzhen Statistical Yearbook 2018. Available online: http://tjj.sz.gov.cn/xxgk/zfxxgkml/tjsj/tjnj/201812/P02018122963972485550.pdf (accessed on 29 December 2018).
39. Shenzhen Internet Rental Bicycle Standard Management Rectification Action Implementation Plan Introduced. Available online: http://jtys.sz.gov.cn/jtzx/wycx/dccx/cxtx/201709/t20170920_8767870.htm (accessed on 20 September 2017).
40. Qi, D.; Li, G.; Tao, Y. Passenger transportation in multi-center city Shenzhen. *Urban Transp. China* **2015**, *2*, 26–33. [CrossRef]
41. Mobike Global. Bike-Sharing and the City 2017 White Paper. Available online: https://mobike.com/global/public/Mobike%20-%20White%20Paper%202017_EN.pdf (accessed on 19 May 2017).
42. Zhang, Y.; Gu, Z.; Zhou, W. Spatial Match of Jobs-housing and the Rail Transit's Role in Commuting Metropolis: A Case Study of Shenzhen. *Urban Plan. Forum* **2018**, *1*, 99–106. [CrossRef]

International Journal of
Geo-Information

Article

An Occupancy Simulator for a Smart Parking System: Developmental Design and Experimental Considerations

Germán Martín Mendoza-Silva *, Michael Gould , Raul Montoliu and Joaquín Torres-Sospedra and Joaquín Huerta

Institute of New Imaging Technologies, Universitat Jaume I, Avda. Vicente Sos Baynat S/N, 12071 Castellón, Spain; gould@uji.es (M.G.); montoliu@uji.es (R.M.); jtorres@uji.es (J.T.-S.); huerta@uji.es (J.H.)
* Correspondence: gmendoza@uji.es

Received: 2 April 2019; Accepted: 3 May 2019; Published: 7 May 2019

Abstract: This paper presents the development of a parking occupancy simulator to support a smart parking system. The simulator uses an agent-based approach to model drivers who follow activity plans and who may or may not use the smart parking system. We illustrate how the process of developing our simulator helped in the design and implementation of the smart parking system components. The paper also shows how the simulator was used to study the possible usage of the smart parking system in a university campus, foreseeing (1) support for the smart parking system's overall suitability, (2) reservation guarantee violation problems, and (3) the value of using total traveled distance as a metric for the smart parking evaluation. The experience presented in this paper may prove valuable to teams planning the development of a smart parking system for similar contexts.

Keywords: on-street parking simulators; smart parking systems; agent-based modeling

1. Introduction

Agent-based simulations have proven valuable for studying traffic and mobility phenomena, including parking search and availability, without disrupting the actual traffic in a city. Many cities assign great importance to solutions to parking-related problems [1]. Those solutions include smart parking systems (SPS) which, mainly found as Parking Guidance and Information (PGI) systems, determine the parking occupancy and provide suggestions about parking availability [2]. SPS have been shown to have positive effects on driver parking success and on traffic flow [3]. SPS development has also significantly benefited from behavior models for parking search that help in analyzing the underlying phenomenon [4], and in testing of design and implementation choices [5,6]. Drivers are commonly represented as agents, using Agent-Based Modeling (ABM) [7,8], located in an urban environment. The environment can be represented using data from Geographic Information Systems (GIS), which provide a digital representation of the urban environment [9] and are already in place in many city governments.

When off-the-shelf simulation software is used to build a parking simulation linked to an SPS, the software often can impose data format and scripting restrictions that create conflicts with the SPS design and implementation. These conflicts prevent or complicate the SPS and simulator from sharing GIS data and algorithm implementation [5,9], affecting resource sharing and thus reutilization. Building or adapting a parking simulator so that it can share data and software components with its related SPS can benefit an SPS project beyond what is explored in previous parking simulations studies. Such benefits are important considering the increasing availability of city geospatial data [10] and the increase in government interest in parking optimization [11]. For example, an SPS development project for a city may benefit from (1) the incorporation of available geospatial services into the SPS development, and (2) the early evaluation of the SPS design and feasibility by means of a parking occupancy simulation.

This paper describes our experiences in building an agent-based parking occupancy simulator. The simulator had two major goals: (1) testing suitability of an SPS in the context of a university campus and (2) reuse of its development efforts (and code) during the SPS development. The SPS targeted by our simulator checks the occupancy of on-street parking spots using sensors, and handles logical spot reservation upon request. The simulator allows the exploration of parking occupancy patterns created by agents that either use or decline to use the SPS. Agents represent drivers of the most typical profiles of people who drive to and within the campus.

As a case study, experimentation was performed using the simulator to explore situations with different levels of parking demand and SPS usage. The results provide insights into a metric for SPS suitability evaluation from a driver's point of view. Also, the experiments allowed exploration of the reservation guarantee problem (someone stealing your assigned spot while you are en route to it), which arises due to the lack of a physical reservation enforcement. In summary, the main experiences and recommendations in this paper are:

1. The methodology we employed to increase re-usability of software development efforts for a parking simulator, applied to a related SPS development;
2. how to explore the reservation guarantee concept for an SPS without physical reservation enforcement; and
3. how to use the total driving distance metric for making credible comparisons when evaluating an SPS usage benefits.

To the best of our knowledge, no previous study has proposed the mentioned reutilization methodology relating an SPS and a parking occupancy simulator. The design proposal allows novel traits like running a simulation from current parking state data, or automatically using the latest environment information. Likewise, despite the fact that the reservation guarantee problem has been acknowledged by other studies, they did not study the problem incidence under several levels of SPS usage. Our analysis questions the acceptability of an SPS that promises a reservation to drivers and does not physically enforce the reservation. Furthermore, the parking studies mainly analyze parking search distance, which is only a part of the total driving distance. Additionally, the code of our simulator is freely available in a public repository.

The remaining sections of this article are as follows. Section 2 presents relevant previous studies and supports our design considerations. Section 3 describes the relationship between the SPS and the parking simulator. Section 4 explains the simulator's details. Section 5 describes the case study of the simulator for the experimental evaluation of SPS usage. Finally, conclusions and acknowledgement sections are presented.

2. Background

This section demonstrates how our design approach is unique by reviewing previous studies. We also review previous SPS evaluation metrics to highlight the relevance of our proposed metrics.

2.1. Smart Parking Systems and Parking Simulations

Modeling and simulation have been used for knowledge discovery applicable to parking systems, as well as for testing already implemented SPS. Some examples of the first approach are: testing a utility function that involves factors affecting parking choice [4], collaborative path-finding in a multi-agent context applied to an SPS [12], testing a parking planning algorithm [13,14], and an SPS evaluation considering several vehicle categories [15]. Examples of the second approach include: testing an SPS model to explore factors like distance to building entrances [16], testing dynamic prices assignment [5], and parking guidance evaluation [6].

Despite the fact that modeling and simulation techniques are often related to SPS design or evaluation [2,17–21], to the best of our knowledge efforts devoted to the simulator/simulation's development are not reutilized in SPS development. A simulator and its related SPS are generally built following distinct goals: The former's development commonly seeks a fast way to study the parking phenomenon, while the latter's development pays more attention to common software concerns like robustness, efficiency and load handling. During our work, we noticed an interesting opportunity to reuse the simulation software modules in a related SPS by combining and linking the simulator development with the development of the guidance/reservation algorithm and other software components for the SPS. The abundance and variety of available agent-based modeling toolkits [17,22] may facilitate such co-developments. The algorithms or other components required in the SPS may be implemented using the same programming tools in the simulator. In a general sense, doing so may require:

1. building or adapting parking simulator software and not just defining a model to run in available modelers,
2. working alongside the SPS development team, and
3. applying software design techniques that assure robust re-usability.

The team that developed the simulator described in this paper was also part of the team developing the targeted SPS. Its members had software design and team skills that enabled them to meet the previous requirements. The software components reutilized in this work are the parking reservation component and the components for accessing the related data and external services. Section 3 presents the selected design decisions that assured the sought-after component reutilization.

2.2. Agent-Based Parking Models And Gis

ABM applications to traffic and transportation, including parking-related phenomena, are significant and numerous [7,8], with several popular ABM toolkits being spatially explicit [23] or including extensions that provide support for GIS data usage [22]. Agents represent drivers in cars moving across an environment, which is usually composed of a network of road, target destinations, and parking spaces.

In research literature related to parking studies, the bridging of ABM and GIS is addressed either using particular spatial data formats or by having the simulation built within a GIS platform. Works like SUSTAPARK [24], TRANSIMS [25], MATSim [26], PARKGRID [27] and PARKAGENT [13,28,29] are good examples. SUSTAPARK and PARKGRID load the roads and parking data from GIS layers stored in files, e.g., in shapefile format. TRANSIMS and MATSim read their input data—e.g., network, destinations, and activity plans, from files that follow their own specification, though they include some GIS tools for importing, exporting, or visualizing other formats. PARKAGENT was implemented as an ArcGIS© application so as to have direct access to GIS data and services.

Following the reutilization goal presented in Section 2.1, we decided to assure the GIS data and services were accessible online and decoupled from the simulator. This decision allowed reutilization of data access, as well as software components. A GIS server providing data access through web services enabled data sharing between several applications, and more specifically, between the simulator and the SPS, thus allowing interesting new considerations—such as running a simulation from the actual parking state detected by the SPS. Our proposal includes the parking spot information and the car and pedestrian route determination hosted as services in a GIS server. The ability to run a simulation from current parking state data, and to automatically use the latest environment information, which is provided by a GIS server, is a distinctive and novel trait of our methodology proposal.

2.3. SPS Evaluation Metrics

The metrics used for evaluating the benefits of using a Smart Parking System include parking search time [5,30], mean driving distance [31], wandering ratio [5], walking time [32], and travel density and average speed [33]. Measurements (most often distance or time) usually start when a car enters the simulated area [31], when it is near to its destination or makes a request for parking [5], or when it arrives to a parking lot [30]. In our model, agents using the SPS are guided as soon as they enter the walled university campus, a city surrogate. The use of total driving distance to achieve credible comparisons of SPS usage benefits is not recommended, given that drivers travel certain distances to their destination regardless of their SPS usage. The total distance is generally valuable, however, and simple to calculate; therefore, we devised a methodology to use the total driving distance in our experiments. Section 5 shows how we arrived at the methodology through a series of experiments using our simulator.

In our model, the driver's travel distance while trying to park might be affected not only by the driver's parking choice, destination, and parking availability. An agent that does not use our SPS may occupy a parking spot already reserved for another agent, forcing the latter to issue a new spot reservation. The effects of reservation without a physical guarantee, i.e., nothing stopping a driver from grabbing someone's reserved spot, are not explored in the literature we reviewed, though reservation guarantee is a key aspect of SPS [5]. The literature does suggest approaches to enforce the reservation, but they are either relatively expensive (physical barriers) or not fully effective [2,5,32]. Section 5 shows how we explored the reservation guarantee problem, and comments on the negative impact it may have for the SPS' usage. To the best of our knowledge, no previous study has hinted on the level of SPS utilization under which the reservation guarantee problem is the most notorious for a given environment.

3. The SPS and the Parking Simulator

The SPS was created and tested at the campus of the Universitat Jaume I (UJI) in Spain. It is a walled complex and has four vehicle entrances. Its parking spots are free and on-street, with some areas similar to those in a small town neighborhood and other larger areas similar to that near a sporting facility. The SPS has detection sensors, smart parking services, and client applications. The magnetic sensors detect the parking occupancy and deliver that information to the smart parking services through a wireless network. These services, exposed as REST web services, handle occupancy data storage and provide search, reservation, and routing functionalities to a smartphone client application. The application allows visualization of available parking spots, spot reservation, and driver guidance.

Our simulator represents drivers that move across the campus to reach their destinations and park in spots convenient to them. The simulator's design allows the usage of the simulated parking occupancy data as a fake (surrogate) input from the SPS sensors. Therefore, the simulator became a valuable tool for the SPS development team for testing the SPS before deployment of the actual occupancy detection sensors in the university campus. Also, the SPS' parking reservation component was implemented and tested as a part of the simulator. Therefore, any further refinement to it could be easily tested through simulations and later be directly used in the SPS. Figure 1 shows the layered design, after Fowler [34], of our simulator software. The Data layer obtains the necessary information for running the simulation. The Model layer represents the actual model (its implementation is aided by an ABM library) along with the implementation of some parts of the SPS. Finally, the Presentation layer has components that handle the model's output and user interaction. The layers vertically communicate using facades [34].

The GIS data and services were hosted using commercial, off-the-shelf GIS server software (Esri (Esri software company (http://www.esri.com/)) ArcGIS for Server, now ArcGIS Enterprise), which is used by many city governments and it is available to universities via the Esri educational institution license. At UJI, this server already hosted production-ready (properly prepared) GIS data and services regarding the university campus, collected and built for the Smart Campus system [35]. This server provided a common access point to geographical data for the simulation and the SPS. The relevant hosted GIS services for the simulation and the SPS were: (1) parking spots data, (2) building and campus entrance points data, and (3) car and pedestrian routing services on a previously digitized street network. These services are consumed as REST web services. Under this uncoupled schema, any changes to parking spots, buildings, or routes are immediately available to the SPS, any related application, and the simulator.

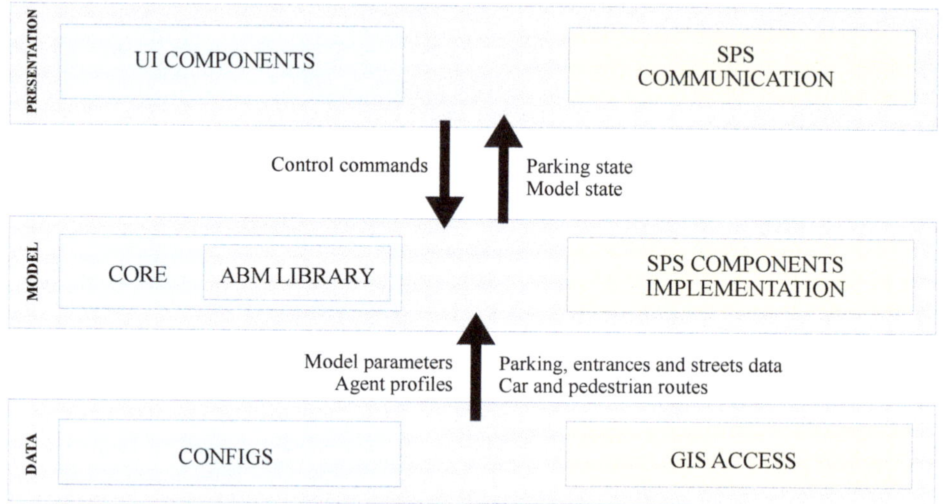

Figure 1. Simulator's layered organization. SPS = smart parking systems; GIS = Geographic Information Systems; ABM = Agent-Based Modeling.

4. Parking Simulator Details

The simulator defines an agent-based model representing the parking occupancy created by drivers within the campus. Its implementation took into account the integration discussed in Section 2.

4.1. The Model

The model represents a 'virtual week' period. For each day of the week, agents arrive to the environment, follow their activity plan, and leave. An agent attempts to park as near to its destination point as possible. Destinations are particular entrance doors of buildings. In studies, such as Geng and Cassandras [5], destinations from the same building are aggregated and considered as one. In our model, given the dispositions of building entrance doors and parking spots, different doors (potentially far apart) of the same building were considered as distinct destinations. The criterion for measuring parking-to-destination proximity is walking distance as measured using actual pedestrian ways (sidewalks and crosswalks).

4.1.1. Agent Profiles

The agent activity plans (agent profiles) define the typical cases of drivers. As an example, consider a student needing to be at a specific classroom at 09:00. At 08:45, she arrives by car to the campus through the campus entrance of her choosing. She then parks near the building door she considers is the best for her destination. After a while, when the class finishes, she walks to her car and drives to the sports complex, which is far from where her car was previously parked. She parks near a door of that complex. After completing her sporting activities she drives out of the campus.

Agents that belong to the same profile have the same type of destinations and similar arrival and departure times. A random, normally distributed variation is allowed around those times. The profiles were created considering the actual number of people from a community profile that commonly visits a building. Agent profiles consider the expected number of people actually driving to the campus for each of those groups. Activity plans are important for transport simulation [36] and the agent profiles addressed in this work resemble those from Horni et al. [26] and Dieussaert et al. [24]. The numbers of people defining each profile and those describing facilities' usage were obtained from the university administration services and community surveys, which is further explained in Section 5. We consider that agent profiles built in this way is a plausible alternative to more common approaches (like car counting at parking spots and at campus entrances) as it requires considerably less effort and infrastructure.

4.1.2. Search Behavior

Some agents ('Guided') rely on the SPS for finding an available parking spot, while other agents ('Explorer') decide for themselves where to park. Studies like Geng and Cassandras [5] have also used these two behaviors, seeking to quantify the benefits from using an SPS. When an agent is created, its type is randomly defined under the restrictions established by a model parameter that controls the proportion between the two types of agents. This model parameter can be dynamically adjusted. Driving behavior and parking search are complex processes [37–39] and several studies have applied realistic behaviors [3,13,19,27], even considering recent trends like driver-less vehicles [40] or specific contexts like a city center [41] and university campuses with specific policies and notable parking supply shortages [42]. We chose two simple parking search behaviors for our model because obtaining an approximate parking occupancy, rather than the most realistic one, was enough for our simulator goal.

Explorer

The 'Explorer' search behavior takes inspiration from Dieussaert et al. [24], Benenson et al. [28], Levy et al. [29], Martens et al. [43], with some simplifications and additions. For example, variations in car speed or maximum search time are not considered, while variable agents' maximum walking distances and two measures for evaluating local parking availability are considered. An 'Explorer' agent first tries to park as close as possible to its destination. If it fails, the agent then tries to park in the first available parking spot it can find. Algorithm 1 presents pseudo-code which briefly describes the Explorer parking search behavior.

Agents move following the shortest network path to their destinations. An agent starts searching for parking when it is within a maximum walking distance to its destination. An agent's maximum walking distance is a random value (*maxWalkDist*) within a range. The agents can detect only parking spaces within a visibility distance, which is defined by a model parameter. While searching, an agent records the proportion of free-to-total parking spots it detects. The agent decides to park in an available spot when (1) the current proportion falls below a critical ratio (*criticalRatio*); or (2) the difference between the proportion from the previous step and the current proportion is greater than a critical value (*criticalReduc*). Both (*criticalRatio*) and (*criticalReduc*) are model parameters.

If an agent has completed the route to its destination without being able to park, it tries to park in the first available parking spot it detects searching first around the target building and then around other buildings. If it does not find available parking places around any building, the agent returns to a campus' exit and leaves.

Algorithm 1: Explorer agents decision rules

Input: *destination, maxWalkDist, criticalRatio, criticalReduc*

found ← **false**
compute shortest *path* to *destination* from current point
while not at the end of *path* **and** *found* = **false do**

  if distance to *destination* ≤ *maxWalkDist* **then**

    pastratio ← *ratio*
    ratio ← local available parking spots / local total parking spots
    diff ← *ratio* − *pastratio*
    if *ratio* > 0 **and** (*ratio* < *criticalRatio* **or** *diff* > *criticalReduc*) **then**

      choose closest reachable parking spot and move to it
      found ← **true**
    end if
  end if
  compute next point in *path* to move to
end while
if *found* = **true then**

  park()
else

  while more destinations to explore **and** *found* = **false do**

    destination ← next entrance of same building or another building
    compute shortest *path* to *destination* from current point
    while not at the end of *path* **and** *found* = **false do**

      get local available parking spots
      if there is available spots **then**

        choose closest reachable parking spot and move to it
        found ← **true**
      end if
      compute next point in *path* to move to
    end while
  end while
  if *found* = **true then**

    park()
  else

    leave the campus
  end if
end if

Guided

An agent with a 'Guided' behavior requests a parking spot reservation from the SPS's reservation component, which chooses the best available spot for its destination. It then follows the optimal route to the spot and occupies it. The reserved spot will not be offered to any other agent until the occupying agent explicitly releases it. As the reservation process is logical, not physical, before a 'Guided' agent arrives to its reserved parking spot, an 'Explorer' agent might find that spot and occupy it. When the 'Guided' agent notices that situation, it asks for a new parking spot and heads for it.

4.2. The Simulator

The chosen ABM library for model formalization and implementation was MASON [44], which facilitated the process of creating an integrated simulation. We additionally used the GeoMASON extension [45], which incorporates support for vector and raster geospatial data. The final implementation followed the software design shown in Figure 1.

The *Data* layer loads the simulation's configuration (including agent profiles) from files, reads the GIS data—which is principally campus cartography—from the GIS Server, and reads parking place status from the SPS. The GIS data for parking spots and entrances are loaded at simulation start, but the routing services are used on-demand. The parking place status information are also loaded at simulation start, thus enabling us to simulate the parking occupancy from a known starting point or from an SPS-provided parking occupancy.

Figure 2 provides an overall, simplified view of the *Model* layer design, which includes model notifications, agents' behavior, model configuration and smart parking artifacts implementation. Any component interested in receiving notifications of model changes must subscribe to the appropriate updater. Distinct agent behaviors are achieved through the basic parking agent specialization. The SPS's reservation component, though it is known and used by the simulator, is implemented independently from the simulation core, thus allowing for easy substitution. By implementing model controllers it is possible to set up the necessary relations, e.g., updaters, according to the target application platform.

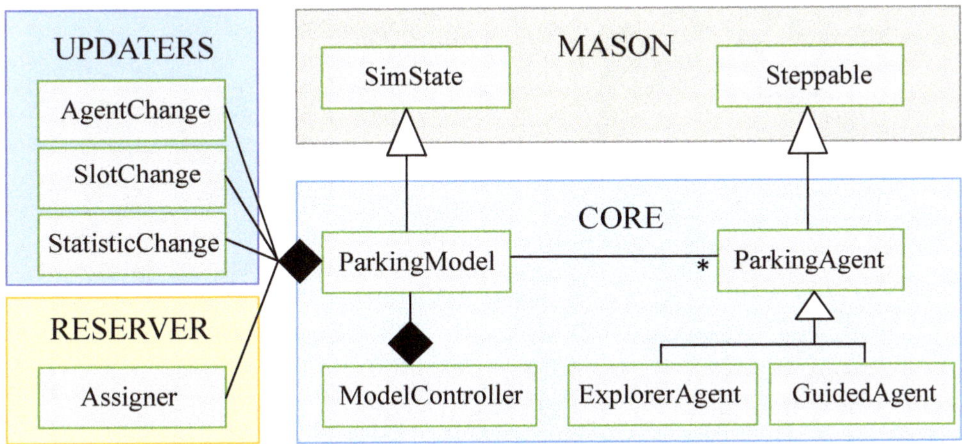

Figure 2. Main classes of the Model layer and their relation.

The *Presentation* layer has components that are notified when the model changes. These components inform the SPS about those changes (Smart Parking Communication) or show them (UI Components) in applications that wrap the simulator. The Smart Parking Communication component delivers occupancy state changes. Three implementations of the simulator's UI Components have resulted in three distinct applications: (1) A desktop version that uses MASON visualization utilities (Figure 3) and shows the parking availability and the agent moving across the campus, (2) a console version for simplified and faster runs, and (3) a web-hosted version. Having three different interfaces for different purposes highlights the flexibility of the approach used for building the simulator. A video showing a simulation run is available online (Demonstration Video: https://youtu.be/gZj21WtOmio).

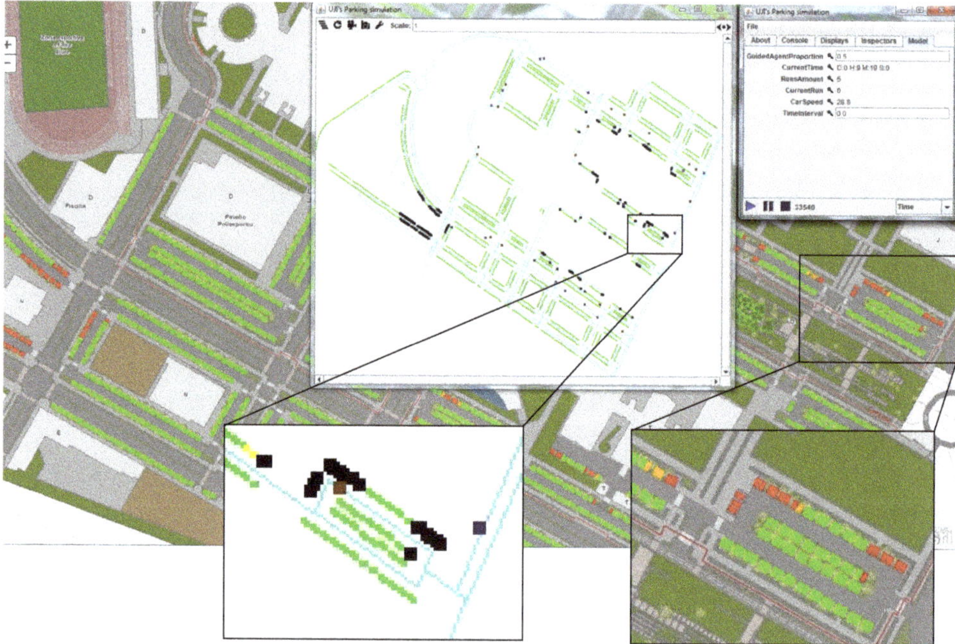

Figure 3. One of the simulator applications running in the foreground, while in the background a web map shows the parking spots' current status as stored in the SPS. In the simulator, black squares represent parked agents. In the web map, red squares represent occupied parking spots.

4.3. Parking Reservation Component

The parking reservation component serves reservation requests in the order they are received. Several studies have proposed advanced reservation algorithms that take into account, e.g., current driver travel time, parking pricing, and reservation updates when the parking availability changes [2,21,32]. As the reservation algorithm was not among this study's main goals, the reservation component uses a simple heuristic that only takes into account the parking spots availability and the walking distance from the spots to the specified destinations. In experimental measurements, network communication lags for car and pedestrian routing requests had medians of 0.43 and 0.08 s, respectively. Car route requests are only made when a car starts moving, thus they do not add a significant burden if made on-demand. An agent's parking reservation request requires determination of the available parking spot that is the closest (according to walking distance) to an agent's destination. To avoid on-demand pedestrian routing requests associated with parking reservation requests, the parking reservation component calculates beforehand, then sorts and stores the distances from every parking spot to every destination. This approach is feasible because the number of destinations (buildings' doors) and parking spots is 225 and 3809, respectively. Therefore, the time for finding the closest available spot only depends on the total number of parking spots.

4.4. Final Development Considerations

The simulator development presented in the previous sections also requires additional minor design and development decisions (All code is shared in a public repository: https://goo.gl/GmWyt1), such as development platform or execution environment. Furthermore, although the usage of resulting simulator software artifacts is straightforward, some additional steps are required if an SPS development is initiated from those artifacts.

The choice of the development platform (and programming language) depends mainly on preferences from the development team and additional design considerations (e.g., related to the SPS design). As can be inferred from Kravari and Bassiliades [17], Crooks et al. [22], it is possible to accommodate model specification (using an ABM library) to the chosen development platform. The usage of an ABM library for model implementation implies coding driver behaviors following the library specifications. For example, using MASON (which is Java-based), driving behavior is coded in methods from a class. An agent is created as an object from that class, and a method from it is executed by a discrete event scheduler. In the case of models considering very complex driving behaviors, the required coding effort may be significant. For those models, a benefit trade-off between re-usability and coding effort should be considered.

The development platform may also influence the way remote services are consumed. The REST architectural style is widely used and its support spans most development platforms. The communication with those services is eased if SDKs exist for the targeted GIS server. For example, the software company Esri provides several runtime SDKs for client applications to access services published by Esri ArcGIS Enterprise, as well as for map visualization and geometrical/geodesic operations. Other SDKs alternative sources could be Mapbox [46] or Geotools [47]. For models that cover large and congested areas (which simultaneously include several thousands of drivers), a large number of network requests are issued, which could potentially overload the GIS server especially if that server already supports multiple users for other purposes. In such situations, a different design (routing operations computed in the simulation) may be preferable.

For the integration of the simulator core components into an application, already mentioned in Section 4.2 for the web or desktop versions, most of the effort is devoted to creating a proper presentation/usage of simulation output and controlling the simulation execution. For example, the web version wraps the simulator core components with REST web services. A thin web client uses them to obtain the simulation state (occupancy) and to display it on a map, and to control the simulation (starting/stopping the simulation and parameters configuration).

Once the simulator is built and tested, the following elements are ready for reuse in an SPS:

1. GIS data and services hosted in a GIS server,
2. software artifacts for remote GIS data read/write operations,
3. reservation component, and
4. visualization components created for simulation testing.

The combination of the enumerated elements with an occupancy detection system (e.g., magnetic sensors and its communication components) and a user (mobile) application can produce a relatively simple, yet functional SPS. As an example, the occupancy or availability state storage, the computation of vehicles routing indications and the maps for visualization are provided by elements in (1). Elements from (2) can be reused to create a web application that provides a gateway for occupancy state discovery and update. The update operations are used by the occupancy detection system. The discovery operations are used by the client application, which also uses elements from (1), and by a web dashboard to monitor the SPS state, which can also reuse elements from (4). The reservation component (3) is vital for the SPS and can be readily used in the SPS as it is isolated from the simulation.

5. Case Study: Exploration of SPS Expected Usage

Several experiments were performed to demonstrate the simulator's potential use and to explore likely benefits of SPS usage. The model parameters' values are presented in Table 1.

Table 1. Model parameters values.

maxWalkDist(m)	[50 to 100]
criticalRatio	0.25
criticalReduc	0.15
visDistance(m)	40
carSpeed(km/h)	30

Model parameters *maxWalkDist*, *criticalRatio*, and *criticalReduc* were already explained in Section 3. Parameter *visDistance* indicates the distance for which 'Explorer' agents consider their detected parking occupancy for deciding whether to park or not. Parameter *carSpeed* sets agents' movement velocity. Agent profile values are not presented here because of their large number of details. The simulator includes other relevant parameters that are not model parameters, for example, the GIS data and routing services, and whether to initialize the simulation with the current SPS occupancy state.

To define destinations, times, and amounts of distinct profiles (groups) of people coming to the campus by car, a set of root behaviors was created based on actual university data for a typical academic month. The root behaviors act as templates used to automatically create agent profiles, which in turn are used as templates for creating agents during a simulation run. The root behaviors used in the experiments considered the main distinct groups of the university members. They defined tasks in which the first drivers arrived to the campus around 08:00 and the last ones left the campus around 20:00. The root behaviors included a main task, which was going to a building, and subsequent optional tasks. Optional tasks were added to agent profiles randomly, complying with the expected facilities usage. The optional tasks were going to a sport facility, to the library, or to a distant building, if the optional task's destination was not already the main task's destination. We obtained the typical quantities of each group accessing each facility and their typical staying times, although for profile creation a group-dependent random variation was applied to staying times. Through driver surveys, we also estimated the likelihood for each group to use a car inside the campus. By creating agent profiles automatically from root profiles, we achieved a rich set of over 300 profiles, which is important in transport simulation [36].

Each experiment consisted of 15 simulation runs with the same parameters. Each experiment was run for a specific proportion p between 'Explorer' and 'Guided'. It also used a profile set created using the root behaviors and considering a specific proportion c of people actually using their car on campus. The maximum amount (N) of people expected to come to the campus on a typical day is about 15,000. Each root behavior (people profile) b accounts for some part N_b of that amount of people. For an experiment run using 0.1 as value for c, the number of agents created for root behavior b is $0.1N_b$, and the expected total amount of agents is (about) 1500. Parameters p and c allow exploration of situations with different levels of SPS usage and parking demand, respectively. The measured total driving distance of each agent considered all stretches of its multi-destination journey. With $D(p,c)$ denoting the set of all traveled distance measurements for a particular experiment, then:

$$D(p,c) = D_G(p,c) \cup D_E(p,c) \tag{1}$$
$$p \in P = \{20\%, 50\%, 80\%\}$$
$$c \in C = \{0.1, 0.2, ..., 1.0\},$$

where $D_G(p,c)$ and $D_E(p,c)$ denote subsets that contain measurements only from 'Guided' agents or from 'Explorer' agents, respectively. We denote the mean values of previous sets as $\overline{D(p,c)}$, $\overline{D_G(p,c)}$, and $\overline{D_E(p,c)}$.

Figure 4 presents the case for $\overline{D(p,c)}$. It shows that the SPS usage should reduce the parking searching time. There is a drop in mean value for 'Guided' agents from proportions 0.1 to 0.2. This may be due to the fact that with proportion 0.1, only a few agents, or no agents at all, have as destinations buildings with a low number of people and located near campus entrances.

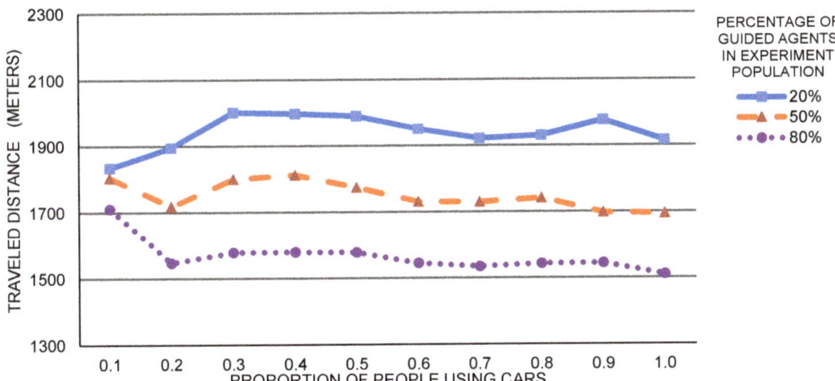

Figure 4. Variation of the mean traveled distance.

The mean traveled distance remained stable in spite of the growing number of drivers. The parking reservation component reserves parking spots that are uniformly distanced from a building target door, causing their mean distance to remain stable for 'Guided' agents. For 'Explorer' agents, we have identified two likely reasons: (1) Parking demand is spread over the day and across several building doors, and (2) when some drivers arrive to campus (most likely after midday), others have already left. Furthermore, parking demand and supply are not heterogeneous across the campus; in a similar way they differ in other scenarios [27], leading to divergences from the expected increase in the parking search time when the parking demand increases. The relation between parking search time and parking demand is not trivial [48].

Figure 5 presents another way to use the mean traveled distance for meaningful comparisons. Each data point $d(p,c)$ of the series is calculated as:

$$d(p,c) = \overline{D_E(p,c)} - \overline{D_G(p,c)}. \tag{2}$$

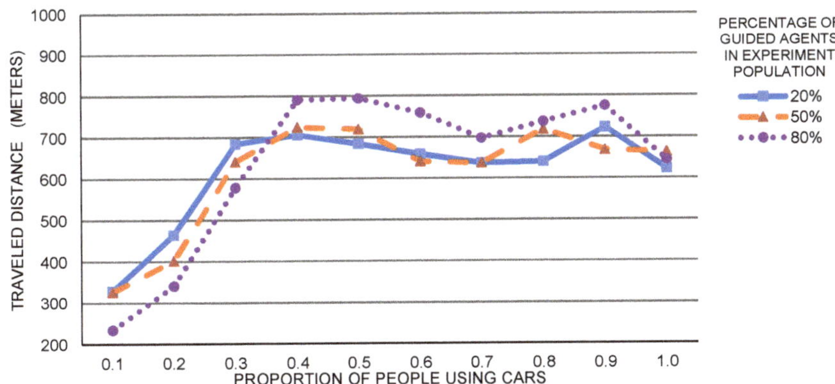

Figure 5. Variation of the mean traveled distance, as expressed by the difference between the mean distance of 'Explorer' and 'Guided' populations.

Figure 5 indicates that the 'Explorer' agents have longer driving distances. The difference stabilizes and stays around 700 m, which may hint at the minimum expected difference in the mean traveled distance in the campus scenario. An unsuccessful parking search across the parking spaces that surround a campus' building would add a value to the accumulated search distance of an agent that, on average, is above 600 m.

When considering all measurements from all experiments, the mean value was 1737 m and the standard deviation was 1174, approximately. The two metrics explored so far in this subsection $(\overline{D_G(p,c)}$ and $\overline{D_E(p,c)}$ and $d(p,c))$ are affected by the distribution of buildings (and their doors), parking spots, and campus' entrances. To avoid that issue, instead of using the mean value, we considered using percentiles values. First, two sets are defined as follows:

$$T_E(p) = \bigcup_{c \in C} \{D_E(p,c), p \in P\} \tag{3}$$

$$T_G(p) = \bigcup_{c \in C} \{D_G(p,c), p \in P\}.$$

Let $n_E(i,p)$ and $n_G(i,p)$ denote the *ith* percentiles of $T_E(p)$ and $T_G(p)$, respectively. The data points of Figure 6 are those percentiles values, considering three levels of usage of the SPS. The chart allows comparisons of distance categories, and it suggests that 'Explorer' agents are strongly affected by high parking demand situations, regardless of the usage level of the SPS. In addition, to compare different levels of usage of the SPS within the same chart, we defined new data points as follows:

$$d(i,p) = n_E(i,p) - n_G(i,p), p \in P. \tag{4}$$

Figure 7 shows the data points $d(i,p)$. The metric value of each proportion $p \in P$ for a percentile are very similar to each other for percentiles below 40, but it significantly differs for the percentiles at about 80. The difference for the highest percentiles is higher for 80% of SPS usage, which is also noticeable in Figure 6. The reason for the higher difference in high SPS usage situations is that 'Guided' agents 'discover' and thus take the remaining spots faster than 'Explorer' agents.

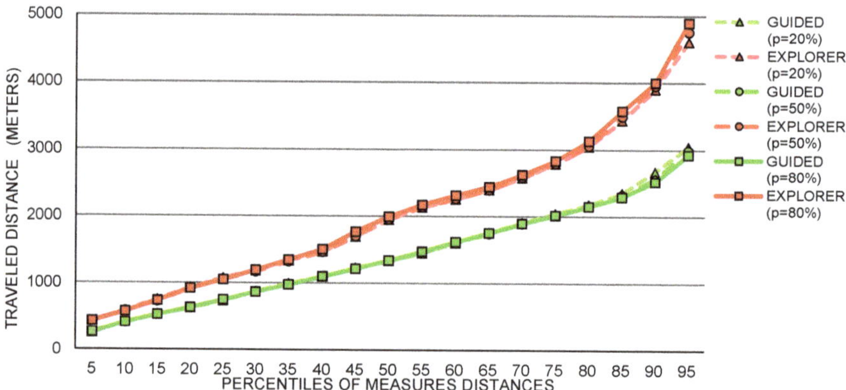

Figure 6. Difference in traveled distance between 'Explorer' and 'Guided' agents, as seen using percentiles and considering $p = 20\%$, $p = 50\%$, and $p = 80\%$.

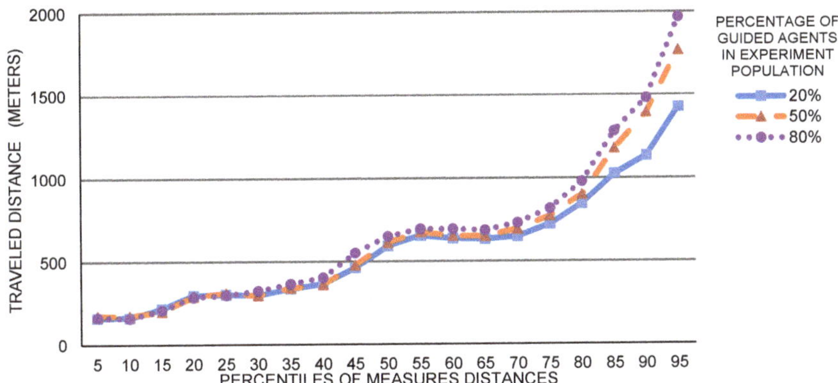

Figure 7. Difference in traveled distance, as expressed by the difference of same percentile values of the 'Explorer' and 'Guided' populations.

Figure 8 shows that the occurrence of the reservation guarantee problem grows as the competition for parking spaces increases. It is most notable when the 'Guided' to 'Explorer' proportion value is 50%. The guarantee problem is more notable for proportion of 75% than 25% because in the former case there are more agents whose reserved spaces are susceptible to be 'stolen'. Drivers declining to use the real SPS may be an issue for those who use it. When a driver has its reserved spot 'stolen' by another driver, the former needs to keep driving to locate an available spot. Depending on the parking demand, such additional search may take a long time. Furthermore, regardless of the amount of time devoted to the additional search, the reservation violation creates discomfort for the driver, which may lead to a decline in SPS usage. The 'reservation' term in an SPS that does not make physical reservation enforcement could be misleading for a driver, and 'suggestion' term should be used instead. When a parking spot is 'stolen' and given the occupancy detection capability of the SPS, a driver that was suggested to take that spot should be alerted and provided with an alternative spot and a new route indication to it. To avoid alternatives far away from the original suggestion, the latter could be determined taken into account the availability of other parking spots close to it, an idea that resembles heuristics applied in PGI systems [2].

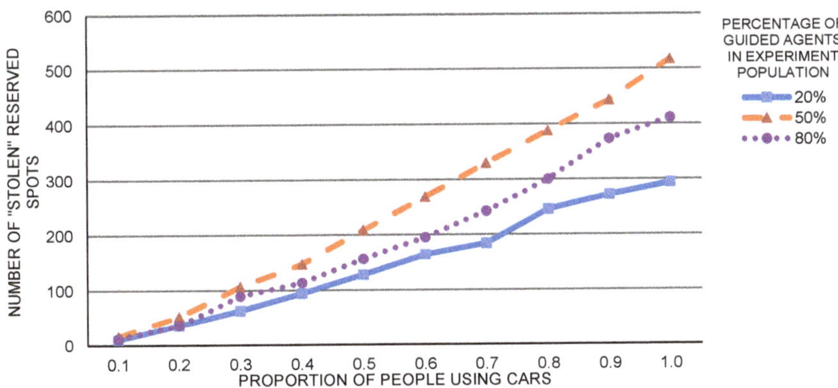

Figure 8. The reservation guarantee problem, explored for several levels of parking demand and agent types proportions.

Single Building Analysis

To study a specific high parking demand situation, e.g., in the case of a notorious event hosted in a campus facility, we ran simulations for which agents tried to park near a randomly chosen entrance of a particular building during a 'virtual' day period. We made them arrive at the campus at similar times. The parking demand significantly exceeded the parking supply around the building, and the agents had to park around other buildings. As presented in Figure 9, in the case of the 'Explorer' agents, the worst result corresponds to the proportion of 'Guided' agents $p = 20\%$, which is a result not only from competition, but also from the fact that the 'Explorer' agents that try to park near the same building's entrance follow similar routes while searching for parking, which makes them go through the least favorable paths. In the case of the 'Guided' agents, the worst results correspond to the proportion $p = 50\%$, which corresponds to when they are the most affected by the reservation guarantee problem.

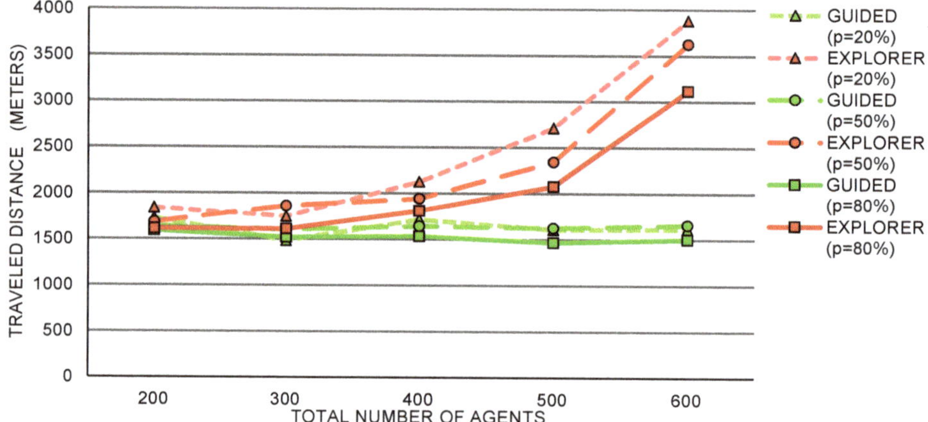

Figure 9. Traveled distance when considering only one campus building.

The correspondence between larger traveled distance and higher parking demand for the 'Explorer' agents is in line with the results presented in previous studies [27]. Likewise, the steady behavior for 'Guided' agents corresponds to already reported benefits of parking reservation systems [2]. However, the comparison between the three studied levels of SPS usage for 'Explorer' agents reveals insights different to those obtained when considering the simulations campus-wide during an entire week. The difference between the two study cases is an indication that the traveled distance does not only depend on the demand and level of usage of the SPS, but also on the scenario. The campus traveled distance analysis results from this work should be applied with caution to other distinct scenarios. The campus has a large amount of parking spots, while in a different scenario, e.g., a city center, the number of parking spots may be rather small. Depending on the scenario, the analysis might have to be performed on small extents. The reservation guarantee problem is, however, inherent to on-street parking, and it should be expected to be the most significant when the numbers of people using and refusing to use an SPS are similar.

A simple driver model and a simple parking reservation heuristic were used in this work to show that an SPS development team could be able to assume the simulation development without significant effort. Having agents with parking search behaviors more advanced than those used in this work would decrease the traveled distance numbers in the campus scenario, e.g., using learning to avoid places that are usually crowded. However, we consider that more general results, like the 'Guided' agents having lower traveled distances than the 'Explorer' agents, and the reservation problem being the most significant when the number of the two agent types are similar, should remain true even when a new a parking search behavior is considered.

6. Conclusions

This paper presented practical considerations from the development of a parking occupancy simulator. The simulator helped in assessing suitability of an SPS suitability for a university campus, and its design and development contributed to reduce the SPS development efforts. The simulator shares software components, GIS data, and services with the SPS. The paper commented on design decisions made regarding the SPS and the parking occupancy simulator, which may be used in similar contexts to minimize development efforts. The application of the proposed methodology on the simulator development will produce software components readily usable in an SPS, as well as GIS data and services readily available online when published on a GIS sever. The software created for this work is available in a public repository, for further technical details inspection and to encourage reproducibility. The paper also presented experimental evaluation of SPS usage benefits using the simulator. The experiments differentiated profiles of agents that use the SPS and those who decline to use it. Analysis of experimental results showed how to use the total driving distance as a metric for evaluating the SPS benefits from a driver point of view. The experimental results also allowed us to explore the effects of having a parking reservation that is logical but not physical, clearly showing that the drivers who decline to use the SPS may "steal" a significant number of already reserved parking spot, which in our experiments reached numbers higher than 100 for a parking demands above the 40 percent of the maximum demand. The guarantee problem becomes more significant as the level of usage of the SPS increases, topping out at 50%, which may hamper drivers' acceptability towards an SPS.

The parking choices used in this work are simple, as they were not among its main goals. Future improvement directions could include the incorporation of more realistic choices, driving behavior, and the ability to learn from experience, which may be combined with an improved reservation algorithm in order to provide agents with updates upon changes in the occupancy state, e.g., when other alternative spots become available. Additionally, a comprehensive model validation with a study case for a different scenario and parking occupancy measurements taken in the field is planned as future work. The design and development guides and experimental analysis presented in this paper show the convenience of including simulations when considering the application of an SPS to a scenario, while diminishing possible dissuasive aspects like model complexity or simulation development efforts, that SPS development teams may consider.

Author Contributions: Conceptualization: Germán M. Mendoza-Silva and Raul Montoliu; Methodology: Germán M. Mendoza-Silva, Michael Gould and Joaquín Torres-Sospedra; Investigation: Germán M. Mendoza-Silva; Software, Germán M. Mendoza-Silva; Supervision: Raul Montoliu andMichael Gould; Resources, Joaquín Torres-Sospedra and Joaquín Huerta; Writing—original draft, Germán M. Mendoza-Silva and Michael Gould; Writing—review & editing, Michael Gould, Joaquín Torres-Sospedra and Joaquín Huerta

Funding: We thank funding from the Spanish' Ministerio de Economía y Competitividad under the project 'SmartWays' (Convocatoria Retos-Colaboración, RTC-2014-1466-4). Germán M. Mendoza-Silva gratefully acknowledges funding from grant PREDOC/2016/55 by Universitat Jaume I.

Acknowledgments: We thank our university's Management Department for providing campus usage and community information.

Conflicts of Interest: No potential conflict of interest was reported by the authors. The funders had no role in the design of the study; in the collection, analyses, or interpretation of data; in the writing of the manuscript, or in the decision to publish the results.

References

1. Inci, E. A review of the economics of parking. *Econ. Transp.* **2015**, *4*, 50–63. [CrossRef]
2. Kotb, A.O.; Shen, Y.C.; Huang, Y. Smart Parking Guidance, Monitoring and Reservations: A Review. *IEEE Intell. Transp. Syst. Mag.* **2017**, *9*, 6–16. [CrossRef]

3. Małecki, K. A computer simulation of traffic flow with on-street parking and drivers' behaviour based on cellular automata and a multi-agent system. *J. Comput. Sci.* **2018**, *28*, 32–42. [CrossRef]

4. Waraich, R.A.; Axhausen, K.W. Agent-based parking choice model. *Transp. Res. Rec.* **2012**, *2319*, 39–46. [CrossRef]

5. Geng, Y.; Cassandras, C.G. New 'smart parking' system based on resource allocation and reservations. *IEEE Trans. Intell. Transp. Syst.* **2013**, *14*, 1129–1139. [CrossRef]

6. Shin, J.H.; Jun, H.B. A study on smart parking guidance algorithm. *Transp. Res. Part C Emerg. Technol.* **2014**, *44*, 299–317. [CrossRef]

7. Bazzan, A.L.C.; Klügl, F. A review on agent-based technology for traffic and transportation. *Knowl. Eng. Rev.* **2014**, *29*, 375–403. [CrossRef]

8. Maggi, E.; Vallino, E. Understanding urban mobility and the impact of public policies: The role of the agent-based models. *Res. Transp. Econ.* **2016**, *55*, 50–59. [CrossRef]

9. Crooks, A.T.; Castle, C.J.E. The integration of agent-based modelling and geographical information for geospatial simulation. In *Agent-Based Models of Geographical Systems*; Springer: Dordrecht, The Netherlands, 2012; pp. 219–251.

10. Dangermond, J. Geospatial Technology and the Future of the City, 2015. Available online: http://www.esri.com/esri-news/arcnews/winter1415articles/geospatial-technology-and-the-future-of-the-city (accessed on 23 April 2019).

11. Frost&Sullivan. Smart Parking to Enable Intelligent Mobility in Global Mega Cities, 2015. Available online: http://ww2.frost.com/news/press-releases/smart-parking-enable-intelligent-mobility-global-mega-cities/ (accessed on 23 April 2019).

12. Rhodes, C.; Blewitt, W.; Sharp, C.; Ushaw, G.; Morgan, G. Smart Routing: A Novel Application of Collaborative Path-Finding to Smart Parking Systems. In Proceedings of the 2014 IEEE 16th Conference on Business Informatics, Geneva, Switzerland, 4–17 July 2014; pp. 119–126.

13. Levy, N.; Render, M.; Benenson, I. Spatially explicit modeling of parking search as a tool for urban parking facilities and policy assessment. *Transp. Policy* **2015**, *39*, 9–20. [CrossRef]

14. Mei, Z.; Feng, C.; Ding, W.; Zhang, L.; Wang, D. Better lucky than rich? Comparative analysis of parking reservation and parking charge. *Transp. Policy* **2019**, *75*, 47–56. [CrossRef]

15. Leclercq, L.; Sénécat, A.; Mariotte, G. Dynamic macroscopic simulation of on-street parking search: A trip-based approach. *Transp. Res. Part B Methodol.* **2017**, *101*, 268–282. [CrossRef]

16. Leephakpreeda, T. Car-parking guidance with fuzzy knowledge-based decision making. *Build. Environ.* **2007**, *42*, 803–809. [CrossRef]

17. Kravari, K.; Bassiliades, N. A survey of agent platforms. *J. Artif. Soc. Soc. Simul.* **2015**, *18*, 11. [CrossRef]

18. Boudali, I.; Ouada, M.B. Smart Parking Reservation System Based on Distributed Multicriteria Approach. *Appl. Artif. Intell.* **2017**, *31*, 518–537. [CrossRef]

19. Ni, X.Y.; Sun, D.J. Agent-Based Modelling and Simulation to Assess the Impact of Parking Reservation System. *J. Adv. Transp.* **2017**, *2017*, 2576094. [CrossRef]

20. Chen, Z.; Spana, S.; Yin, Y.; Du, Y. An Advanced Parking Navigation System for Downtown Parking. *Netw. Spat. Econ.* **2019**. doi:10.1007/s11067-019-9443-4. [CrossRef]

21. Lin, T.; Rivano, H.; Le Mouël, F. A survey of smart parking solutions. *IEEE Trans. Intell. Transp. Syst.* **2017**, *18*, 3229–3253. [CrossRef]

22. Crooks, A.; Malleson, N.; Manley, E.; Heppenstall, A. *Agent-Based Modelling and Geographical Information Systems: A Practical Primer*; SAGE Publications Limited: London, UK, 2018.

23. Taillandier, P.; Gaudou, B.; Grignard, A.; Huynh, Q.N.; Marilleau, N.; Caillou, P.; Philippon, D.; Drogoul, A. Building, composing and experimenting complex spatial models with the GAMA platform. *GeoInformatica* **2018**. doi:10.1007/s10707-018-00339-6. [CrossRef]

24. Dieussaert, K.; Aerts, K.; Steenberghen, T.; Maerivoet, S.; Spitaels, K. SUSTAPARK: An agent-based model for simulating parking search. In Proceedings of the AGILE International Conference on Geographic Information Science, Hannover, Germany, 2–5 June 2009.

25. Lee, K.S.; Eom, J.K.; seop Moon, D. Applications of TRANSIMS in Transportation: A Literature Review. *Procedia Comput. Sci.* **2014**, *32*, 769–773. [CrossRef]

26. Horni, A.; Nagel, K.; Axhausen, K.W. *The Multi-Agent Transport Simulation MATSim*; Ubiquity Press: London, UK, 2016.

27. Fulman, N.; Benenson, I. Agent-Based Modeling for Transportation Planning: A Method for Estimating Parking Search Time Based on Demand and Supply. *arXiv* **2018**, arXiv:1806.10874.

28. Benenson, I.; Martens, K.; Birfir, S. PARKAGENT: An agent-based model of parking in the city. *Comput. Environ. Urban Syst.* **2008**, *32*, 431–439. [CrossRef]

29. Levy, N.; Martens, K.; Benenson, I. Exploring cruising using agent-based and analytical models of parking. *Transp. A Transp. Sci.* **2013**, *9*, 773–797. [CrossRef]

30. Surpris, G.; Liu, D.; Vincenzi, D. How Much Can a Smart Parking System Save You? *Ergon. Des. Q. Hum. Factors Appl.* **2014**, *22*, 15–20. [CrossRef]

31. Wang, H.; He, W. A Reservation-based Smart Parking System. In Proceedings of the 2011 IEEE Conference on Computer Communications Workshops (INFOCOM WKSHPS), Shanghai, China, 10–15 April 2011; pp. 690–695.

32. Tasseron, G.; Martens, K. Urban parking space reservation through bottom-up information provision: An agent-based analysis. *Comput. Environ. Urban Syst.* **2017**, *64*, 30–41. [CrossRef]

33. Cao, J.; Menendez, M. System dynamics of urban traffic based on its parking-related-states. *Transp. Res. Part B Methodol.* **2015**, *81*, 718–736. [CrossRef]

34. Fowler, M. *Patterns of Enterprise Application Architecture*; Addison-Wesley Longman Publishing Co., Inc.: Reading, MA, USA, 2002.

35. Torres-Sospedra, J.; Avariento, J.; Rambla, D.; Montoliu, R.; Casteleyn, S.; Benedito-Bordonau, M.; Gould, M.; Huerta, J. Enhancing integrated indoor/outdoor mobility in a smart campus. *Int. J. Geogr. Inf. Sci.* **2015**, *29*, 1955–1968. [CrossRef]

36. Baqueri, S.F.A.; Adnan, M.; Kochan, B.; Bellemans, T. Activity-based model for medium-sized cities considering external activity–travel: Enhancing FEATHERS framework. *Future Gener. Comput. Syst.* **2019**, *96*, 51–63. [CrossRef]

37. Chaniotakis, E.; Pel, A.J. Drivers' parking location choice under uncertain parking availability and search times: A stated preference experiment. *Transp. Res. Part A Policy Pract.* **2015**, *82*, 228–239. [CrossRef]

38. Zhao, C.; Li, S.; Wang, W.; Li, X.; Du, Y. Advanced Parking Space Management Strategy Design: An Agent-Based Simulation Optimization Approach. *Transp. Res. Rec.* **2018**. [CrossRef]

39. Antolín, G.; Ibeas, Á.; Alonso, B.; dell'Olio, L. Modelling parking behaviour considering users heterogeneities. *Transp. Policy* **2018**, *67*, 23–30. [CrossRef]

40. Bischoff, J.; Maciejewski, M.; Schlenther, T.; Nagel, K. Autonomous vehicles and their impact on parking search. *IEEE Intell. Transp. Syst. Mag.* **2018**. [CrossRef]

41. Khaliq, A.; van der Waerden, P.; Janssens, D.; Wets, G. A Conceptual Framework for Forecasting Car Driver's On-Street Parking Decisions. *Transp. Res. Procedia* **2019**, *37*, 131–138. [CrossRef]

42. Meng, F.; Du, Y.; Li, Y.C.; Wong, S.C. Modeling heterogeneous parking choice behavior on university campuses. *Transp. Plan. Technol.* **2018**, *41*, 154–169. [CrossRef]

43. Martens, K.; Benenson, I.; Levy, N. The dilemma of on-street parking policy: Exploring cruising for parking using an agent-based model. In *Geospatial Analysis and Modelling of Urban Structure and Dynamics*; Springer: Berlin, Germany, 2010; pp. 121–138.

44. Luke, S. Multiagent simulation and the MASON library, 2015. Available online: https://cs.gmu.edu/~eclab/projects/mason/manual.pdf (accessed on 23 April 2019).

45. Coletti, M. The GeoMason Cookbook, 2013. Available online: https://cs.gmu.edu/~eclab/projects/mason/extensions/geomason/geomason.pdf (accessed on 23 April 2019).

46. Kastanakis, B. *Mapbox Cookbook*; Packt Publishing Ltd.: Birmingham, UK, 2016.

47. GeoTools: The Open Source Java GIS Toolkit, 2019. Available online: https://geotools.org/ (accessed on 23 April 2019).

48. Arnott, R.; Williams, P. Cruising for parking around a circle. *Transp. Res. Part B Methodol.* **2017**, *104*, 357–375. [CrossRef]

MDPI
St. Alban-Anlage 66
4052 Basel
Switzerland
Tel. +41 61 683 77 34
Fax +41 61 302 89 18
www.mdpi.com

ISPRS International Journal of Geo-Information Editorial Office
E-mail: ijgi@mdpi.com
www.mdpi.com/journal/ijgi